Digital Transmission

ESSEX SERIES IN TELECOMMUNICATION AND INFORMATION SYSTEMS

Series editors
Andy Downton
Edwin Jones

Forthcoming Titles
Satellite and Mobile Radio Systems
Real-time Computing for Telecommunication Systems
Coherent Optical Communication Systems

Other Titles
Computer Communication Networks
Engineering the Human–Computer Interface
Image Processing
Speech Processing

DIGITAL TRANSMISSION

Edited by

Edwin Jones

Department of Electronic Systems Engineering
University of Essex

McGRAW-HILL BOOK COMPANY

London · New York · St Louis · San Francisco · Auckland
Bogotá · Caracas · Lisbon · Madrid · Mexico · Milan · Montreal
New Delhi · Panama · Paris · San Juan · São Paulo
Singapore · Sydney · Tokyo · Toronto

Published by
McGRAW-HILL Book Company Europe
SHOPPENHANGERS ROAD · MAIDENHEAD · BERKSHIRE · SL6 2QL · ENGLAND
TEL: 0628 23432
FAX: 0628 770224

British Library Cataloguing in Publication Data
Digital Transmission. – (Essex Series in
Telecommunication & Information Systems)
 I. Jones, Edwin II. Series
 621.382

 ISBN 0–07–707810–1

Library of Congress Cataloging-in-Publication Data
Digital transmission / edited by Edwin Jones.
  p.   cm. –– (Essex series in telecommunication and information
 systems)
  Includes bibliographical references and index.
  ISBN 0–07–707810–1
  1. Digital communications.   2. Integrated services digital
 networks––Equipment and supplies.   I. Jones, Edwin
 II. Series.
 TK5103.7.D544   1993
 621.382––dc20   93–17927
                                                       CIP

1234 CUP 9543

Typeset by Datix International Limited, Bungay, Suffolk
and printed and bound in Great Britain at the University Press, Cambridge

Contents

Part 2 Signal multiplexing and coding   93

Part 3   Systems and networks   201

Notes on contributors

(in chapter order)

Edwin Jones studied digital transmission systems for his PhD which he received in 1974. He is a senior lecturer in Telecommunication Systems at the University of Essex, having formerly been with GEC Research Limited. He has particular teaching and research interests in the field of digital transmission and has organized the short course on this topic at Essex since its inception in 1987. Current research projects include: efficient speech and data coding for local area networks, signal processing and coding for high capacity transmission systems, and techniques for data synchronization. Much of this is collaborative work with the UK telecommunications industry.

Shamim Siddiqui lectures in Optical Communication Systems. His current research interests include polarization effects in optical fibre communication systems, optical fibre amplifiers, and coherent detection techniques.

Larry Lind was awarded a PhD from Leeds University in 1968. He is currently Head of the Analogue and Digital Signal Techniques Research Group in the Department of Electronic Systems Engineering at the University of Essex. His research interests are in filter theory and data transmission, and he has published over 90 papers in these areas. He is a Senior Member of the IEEE and Chairman of the IEEE Region 8 Circuits and Systems Chapter. He is also a Fellow of the IEE and has served on the IEE Professional Group E10 (Circuit Theory and Design).

Nigel Newton was a postgraduate student at Imperial College, London and received the PhD degree of the University of London in 1984. After working for a year as a Postdoctoral Research Assistant in the same department he joined the Department of Electronic Systems Engineering at the University of Essex as a Lecturer. His research interests include the mathematical aspects of telecommunications engineering.

Steve Whitt has been with British Telecom since graduating from Gonville & Caius College, Cambridge, in 1980. Initially he was involved in research in the field of high-speed digital transmission systems, but more recent responsibilities have included network design and development for the synchronous digital hierarchy. Currently he leads a team conducting strategic studies into network performance and whole life costs.

Jeffrey Deslandes received a PhD from the University of Essex in 1991. He is a research engineer in the Systems Research Division of the British Telecom Laboratories. He has particular interests in the measurement of errors in digital transmission systems and their impact on multiplexers. His current research is into the error propagation characteristics of the synchronous digital hierarchy.

Ken Cattermole is a well-known expert on many facets of telecommunications ranging from the technicalities of transmission and switching through to the social and economic aspects of system provision. He is the author of numerous papers and several books. He became the founding Professor of Telecommunications at the University of Essex in 1968 and now holds the title of Emeritus Professor. In 1992 he was awarded the title of Doctor of the University in recognition of many years' service to the industrial and academic telecommunication community.

Goff Hill graduated from the University of Newcastle upon Tyne in 1969, with an honours degree in Electrical Engineering. Since then he has worked in BT's Laboratories including, for the last eight years, research into optical communications networks. Currently he leads the Advanced Networks Design group which provides support to BT's network planners. Under his leadership, the group has developed the concept of an optical transport network layer.

Mike O'Mahony was awarded a PhD from the University of Essex in 1977. He joined BT Laboratories working on optical transmission systems and became Head of Inland Optical Systems in 1988. In 1991 he joined the Department of Electronic Systems Engineering at Essex as Professor of Communication Systems and Networks and is currently involved in research into optical systems and networks.

Charles Hughes holds a part-time post of Professor of Communications at the University of Essex; he was formerly a Deputy Director at British Telecom Research Laboratories. He has wide industrial experience in both switched networks and radio systems and has published widely in both fields. His current research interests include the study of multiservice asynchronous transfer mode networks and cellular radio systems.

Don Pearson studied for his PhD at Imperial College, London, which he received in 1965. He then spent a period at Bell Laboratories in the

USA, returning to England in 1969 to take up an appointment at the
University of Essex. He founded the Visual Systems Research
Laboratory at the university and in 1986 was appointed Professor of
Telecommunication and Information Systems. He has published widely
in the field of image processing and visual communications and has
been involved in the organization of a number of international
conferences on image communication. Since 1980 he has been a member
of Professional Group E4 Committee (Image Processing and Vision) of
the IEE, serving as Chairman from 1987 to 1989.

Peter Cochrane is a graduate of Trent Polytechnic and the University of
Essex. He is currently a visiting professor to the Universities of Essex,
Southampton and Kent. He joined BT Laboratories in 1973 and has
worked on both analogue and digital switching and transmission studies
at national and international level. From 1988 he managed the Long
Lines Division where he was involved in the development of intensity-
modulated and coherent optical systems, photonic amplifiers, and
wavelength-routed networks. He has published widely in all areas of
transmission system studies. He is currently the Systems Research
Divisional Manager.

Series preface

This book is part of a series, the *Essex Series in Telecommunication and Information Systems*, which has developed from a set of short courses run by the Department of Electronic Systems Engineering at the University of Essex since 1987. The courses are presented as one-week modules on the Department's MSc in Telecommunication and Information Systems, and are offered simultaneously as industry short courses. To date, a total of over 800 industrial personnel have attended the courses, in addition to the 70 or so postgraduate students registered each year for the MSc. The flexibility of the short course format means that the contents both of individual courses and of the courses offered from year to year are able to reflect current industrial and academic demand.

The aim of the book series is to provide readable yet authoritative coverage of key topics within the field of telecommunication and information systems. Being derived from a highly regarded university postgraduate course, the books are well suited to use in advanced taught courses at universities and polytechnics, and as a starting point and background reference for researchers. Equally, the industrial orientation of the courses ensures that both the content and the presentation style are suited to the needs of the professional engineer in mid-career.

The books in the series are based largely on the course notes circulated to students, and so have been 'class-tested' several times before publication. Though primarily authored and edited by academic staff at Essex, where appropriate each book includes chapters contributed by acknowledged experts from other universities, research establishments and industry (originally presented as seminars on the courses). Our colleagues at British Telecom Laboratories, Martlesham, have also provided advice and assistance in developing course syllabuses and ensuring that the material included correctly reflects industry practice as well as academic principles.

As series editors we would like to acknowledge the tremendous support we have had in developing the concept of the series from the original idea through to the publication of the first group of books. The successful completion of this project would not have been possible without the substantial commitment shown not only by individual authors but by the Department of Electronic Systems Engineering as a whole to this project. Particular thanks go to the editors of the individual books, each of whom, in addition to writing several chapters, was responsible for integrating the various contributors' chapters of his or her book into a coherent whole.

October 1992

Andy Downton
Edwin Jones

Introduction

Digital Transmission, the title of this book, will no doubt imply different topics to different readers. At its most obvious, digital transmission means the communication of individual digital (or data) pulses over a communication link from one point to another. However, with modern digital communication systems, point-to-point pulse transmission is only part of a much bigger story. After considering the design and transmission performance of individual pulses, the implications of formatting a user's data into sequences of individual pulses for transmission must also be addressed. Careful sequence coding is required to ensure that the transmitted data stream includes the necessary desirable features for both efficient and reliable transmission. Yet, for many modern applications this is still not enough. Our efficient and reliable links must be connected together to form networks through which the user's information has to pass on its way to the intended destination. This requires the study of a whole variety of systems and networking topics, some of which are still the subject of intense study (and some contention!) within the networking community.

In recent years the authors of this book have been involved in teaching an intensive one-week short course which has attempted to address the above wide range of topics under the general heading of digital transmission. The aim of our course, and also this book, is to bring together both academic and industrial viewpoints on the principles and practice of digital transmission; from pulse transmission, through multiplexing and coding to systems and network design issues. This book has evolved from the set of notes for our course. Using a multi-authored blend of theory and practice, the objective has been to offer a middle course between the two; after all, successful new designs involve a judicious fusion of theory and practical constraints. With the wide range of topics implied above and the constraints of time and space, not least the reader's time for study, our aim is to provide the essential principles and an overall picture of current practices and trends such that the reader has the confidence to follow up further details as necessary in the many references cited. This is not to say that we have avoided all detailed analysis. A number of chapters contain an in-depth treatment of their subject where it is felt that this is essential for an adequate understanding of the topic. Some of the analytical techniques developed in these chapters can also be applied to other aspects of digital transmission engineering.

We consider the three separate but interrelated topics mentioned above by arranging the book into three parts.

Part 1   Pulse transmission principles

The book assumes the reader has at least an elementary appreciation of the fundamentals of *pulse transmission and regeneration.* In Chapter 1 Edwin Jones summarizes these fundamentals by reviewing the basic theory and practice of pulse shaping and bandwidth requirements in the presence of noise. Reference is made to the important theorems of Nyquist and Shannon. Of course, a study of such matters is not complete without a knowledge of the characteristics and limitations of the transmission channel and the equalization of it. These topics are the subjects of the next two chapters. In Chapter 2 Shamim Siddiqui discusses *the optical fibre channel* from the point of view of pulse transmission. This whole chapter has been devoted to optical fibre as it is the subject of so much current development and new investment. The fundamental topics discussed in the first chapter are brought to a focus when channel *equalization* is considered. This is the subject of Chapter 3 by Larry Lind where, using example-based explanations, he discusses various methods of compensation for channel distortions. First frequency domain based techniques and then time domain methods are explored.

Chapter 4 returns to a consideration of channels. By building on the baseband principles reviewed in the first chapter, Larry Lind looks at *modulation for bandpass channels*, a topic where the quest for ever more effective use of signal space is especially evident. Part 1 of the book is concluded with a chapter of a deliberately rather different style; Nigel Newton in Chapter 5 looks at the more theoretical aspects of *making digital decisions.* It deliberately introduces the reader to a more mathematical approach with the aim of making some important points about the objectives and theoretical bounds to the performance of digital detection systems in the presence of noise. Examples are used to illustrate the techniques and the practical limitations of optimal decision-making, culminating in a treatment of the Viterbi algorithm.

Part 2   Signal multiplexing and coding

This part of the book considers the various terminal processing functions that ensure that a digital communication link exhibits the requisite desirable features for efficient and reliable transmission. The relative importance of features such as the efficient use of the channel, or careful monitoring and control of transmission errors, will vary from one application to another. For many applications, a major objective will be channel efficiency, in terms of the bandwidth and the signal-to-noise ratio required for a given information-carrying capacity. This means that data from several sources will often have to be multiplexed together to ensure an adequate composite data channel rate. Thus the first chapter in this part, Chapter 6 by Steve Whitt, discusses *digital multiplexing principles*, including the economics of multiplexing, and provides an introduction to the plesiochronous and synchronous hierarchies used in practice. This culminates in a discussion of the synchronous digital hierarchy (SDH); a topic that is a recurring one in later chapters.

Digital transmission, and digital multiplexing in particular, inserts structure into the transmitted pulse stream, a structure that must be reliably identified at the distant receiver if the user's data is to be satisfactorily recovered. In Chapter 7 Edwin Jones deals with *data synchronization*. He first deals with the extraction of a reliable clock from an incoming data stream and so formalizes the bit synchronization requirements for pulse sequence regeneration assumed in Part 1 of the book. A graphical representation is used to provide an intuitive explanation of the source of clock information and the requirements for minimizing jitter. The second part of this chapter deals with frame synchronization, that is, the insertion of frame-marking information at the transmitter so that the frame structure can be reliably located at the receiving terminal.

Chapter 8 by Jeffrey Deslandes moves on to the subject of *errors and error extension* in digital transmission systems. Ultimately, it is the measure of errors arising (principally) from the noisy and pulse-distorting transmission channel that determines the quality and so the acceptability of a digital communication path. This chapter starts by looking at the various measures that are used to specify the error performance of a transmission link. As always when dealing with such statistical quantities, various measures for specifying, both short and long term, have to be adopted. The chapter then goes on to look at the interesting topic of error extension when demultiplexing. This relates to the previous two chapters on multiplexing and frame synchronization and addresses the question of what happens in the demultiplexer when the frame aligner gets things wrong.

The theme of transmission errors is continued in Chapter 9 where Ken Cattermole presents a broad-ranging treatment of *error control coding*. After introducing the general ideas of error detection and correction together with the necessary mathematics of algebraic structure, he uses selected examples to explain the principles of both block and convolutional coding techniques. The approach, using nontrivial examples and graphical techniques, provides a valuable insight. His explanation of Viterbi decoding reinforces the treatment of convolutional coding and is written to complement the formal probabilistic theory in the context of decision-making in Chapter 5.

Our treatment of coding concludes with Chapter 10 on *line codes* written by Edwin Jones. Line codes are used in most digital transmission systems to ensure that the transmitted pulse sequence has the requisite properties for reliable detection at the receiving terminal. This chapter draws together a number of topics from earlier chapters, in that line coding must include a consideration of transmission bandwidth (and so equalization), choice of code radix (and so decision-making), clock extraction, and the detection of transmission errors.

Part 3   Systems and networks

A challenge for authors who wish to include up-to-date material on digital systems and networks is to present in a limited space a representative treatment of the many facets of this extensive and rapidly changing subject. Here we are only able to present a limited selection of topics; we have chosen to concentrate on telephony-based transmission systems. The reason for doing this is that these are the systems in extensive use worldwide, and while they may not be entirely representative of all digital systems, they provide a framework in which to study many important concepts, practices, and trends.

By way of introduction, Chapter 11 presents a *review of digital networks* in which Goff Hill explains the terminology and takes the reader through the issues which are shaping the future digital communication systems. This chapter sets the scene for the network chapters that follow. Many of the networks of the future are expected to rely heavily on optical fibre; this transmission medium is now well established for long-distance connections and is looking increasingly attractive for shorter distances also. Thus, Chapter 12 by Mike O'Mahony on *optical transmission in telecommunication networks* deals with issues relating to current and future directions of optical fibre transmission in digital networks. Its impact on modern network design is also discussed.

The question of when one should use radio and when cable for transmission is likely to become more central as the demand for radio bandwidth becomes ever greater. However, mobile systems will remain a clear-cut case and they are the subject of Chapter 13 where Charles Hughes explains the important factors to be considered when designing *mobile radio systems.* This chapter brings together a number of topics from earlier chapters and, apart from its value as a study of mobile radio design techniques, it also acts as a tutorial example of how many of the topics dealt with earlier in the book are combined in a modern communication system.

Another topic of current debate is that of the *broadband ISDN and the future services* it is expected to support. Don Pearson tackles this subject in Chapter 14 and considers the issues that network designers

will have to address when developing such services. The chapter provides
a thought-provoking glimpse into possible teleservices of the future that
can offer increased utility and realism. Finally, to conclude this section
and indeed the whole book, Peter Cochrane in Chapter 15 reviews the
development, both past and future, of the digital transmission scene
from copper to glass. This chapter makes a number of interesting
comparisons that support an illuminating review of where digital
transmission has come from, where it is likely to go in the near future
and then, with some extrapolation, imagination, and challenges to
established wisdom, where it might go after that.

Acknowledgements

Our courses have attracted a large number of students from a wide cross-section of industry as well as from academia. All the contributors to this book wish to express their thanks to the many students who, through their enthusiasm and questioning, have helped to mould the content of the course (and so this book) into its present form. As editor and organizer of our *Digital Transmission* course, I would like to thank my colleagues the co-authors who have given lectures and seminars and who have turned their course notes into chapters for this book; it has been a pleasure to work with them. Grateful thanks are also due to Lynne Murrell, Pat Crawford and Lynne Burnand in the Telecommunications Office at the University of Essex, for their cheerful help. Finally as editor, I thank Camilla Myers and Rosalind Hann at McGraw-Hill for their encouragement, advice and, as I write this, patience in generous measure.

August 1993   Edwin Jones

PART I

Pulse transmission principles

1 Pulse transmission and regeneration—a review

EDWIN JONES

1.1 System requirements

A simplified block diagram of a digital communication system is shown in Figure 1.1. Information symbols from a *data source* are processed so that they can be reliably transmitted via a communication *channel* to a distant *data terminal*. The user will judge the quality of the system in terms of the difference between the transmitted data and the received data. The system designer's task is to ensure that this difference is always acceptably low, a task which will nearly always involve a compromise between performance and cost. Performance will be measured in terms of *bit error ratios* over specified time intervals, while cost will be determined by both the channel bandwidth requirements and by the processing complexity in the transmitter and receiver.

These processing tasks fall into two categories:

- *Source coding* is concerned with the structure and statistics of the data source. It will often entail a conversion of the data into a more suitable form for subsequent processing. Ideally, as information theory tells us, source coding should ensure that the information

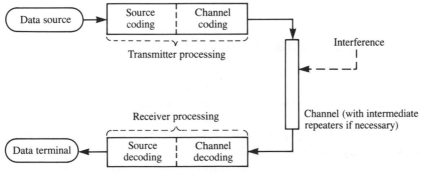

Figure 1.1 A digital communication system

content of a data source is 'smoothed out' so that on transmission the prominence of each data symbol, such as its amplitude or transmission time, is a function of its importance. Word alignment and multiplexing (when several data sources are combined) and the associated task of frame alignment can also be regarded as part of the source coder's task. If the information source is in analogue form (speech or video), then the requisite analogue-to-digital conversion process is also a source coding task.

● *Channel coding* is concerned with the characteristics of the transmission channel; it must ensure that the processed data is compatible with the requirements of the channel. For example, it may be necessary to add redundancy to the source-coded data symbols to combat the effects of transmission errors by providing some error detection and/or correction capability. Symbol timing information may also need to be added to assist the regenerating process at the distant receiver and also at intermediate repeaters if they are necessary. In addition, channel coding must ensure that the frequency spectrum of the processed data is compatible with that of the channel, thus *line coding* for low-pass channels and *modulation* (carrier keying) for band-pass channels are also tasks for the channel coder.

In this chapter we concentrate on the fundamental aspects of transporting individual pulses over a transmission channel. Of course, we must be mindful that there will usually be correlation between neighbouring pulses brought about by source and channel coding of various kinds; these topics will be addressed in later chapters. Now, we highlight the essential principles of pulse transmission by further concentrating on transmission over low-pass channels (i.e. metallic cable). These same principles will also apply to band-pass channels (radio and optical fibres); these topics are also treated in later chapters.

In practice it is not always possible to make a clear distinction between source and channel coding functions. Indeed, improved efficiency can sometimes be obtained by combining some tasks as discussed, for example, in Hamming (1980).

1.2 The baseband waveform

Figure 1.2(a) shows a binary data signal, assumed to be in a suitably coded form for transmission over a low-pass channel, this is often referred to as the *baseband waveform*. After transmission over the restricted bandwidth of a channel (and if necessary, some equalization to compensate partially for the characteristics of the channel) the signal may, typically, look like Fig. 1.2(b).

The task of the receiver is to establish the transmitted sequence. For a binary sequence, a single *decision threshold* is required. Each amplitude

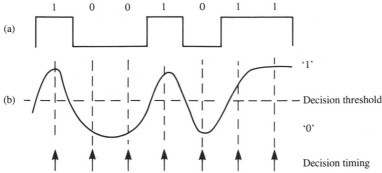

Figure 1.2   Baseband waveforms: (a) transmitted signal; (b) bandlimited received signal

decision will have to be made in the presence of any distortion and interference superimposed upon the data sequence during transmission. The decision-making process will be more reliable if it is timed by a clock which distinguishes when each received symbol has reached an optimum value. The accuracy with which the decision threshold and sample times must be placed will depend upon the severity of distortion suffered during transmission. Wideband channels lead to less critical amplitude and time placement requirements, but they may admit more noise into the receiver. A compromise must therefore be struck, as we shall discuss later.

The interrelated tasks of equalization, clock extraction and decision-making are shown in Fig. 1.3. Also shown is the output stage which, under the control of the extracted clock, produces a retimed baseband waveform. Most digital transmission systems use such a *self-timed fully regenerative* arrangement at the distant terminal. A similar configuration is common at intermediate repeaters (if used), although the simplicity of untimed amplifiers have advantages in some optical systems.

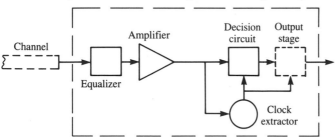

Figure 1.3   A baseband regenerator

1.3   Noise and the decision process

The effect of noise on the decision process at a digital receiver is treated
in many textbooks, for example, Carlson (1986), Sklar (1988) and Lee
and Messerschmitt (1988). For zero-mean Gaussian noise, the
probability of making an erroneous decision is given by:

$$P_e = \frac{1}{\sigma(2\pi)^{\frac{1}{2}}} \int_a^\infty \exp - x^2/2\sigma^2 \, \mathrm{d}x \tag{1.1}$$

where a is the voltage difference between the received signal and its
associated decision threshold at the decision time, and σ is the r.m.s.
noise voltage.

This is known as the *Gaussian tail* area formula (or complementary
error function) and can be shortened to:

$$P_e = \mathrm{tail}\left(\frac{a}{\sigma}\right) \tag{1.2}$$

Unfortunately, there is no analytic solution to this question and so
tables (published as an appendix in many books on digital
communication), bounds or approximation formulae (Beaulieu, 1989)

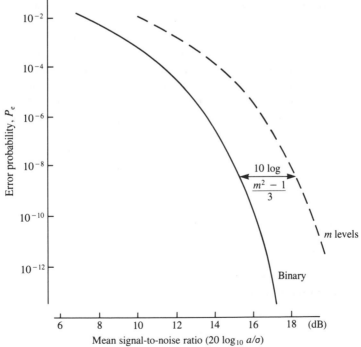

Figure 1.4   Error probability due to Gaussian noise

have to be used. The solid line of Fig. 1.4 shows the result plotted as a function of the signal-to-noise ratio at the input to the decision circuit. This line is also equivalent to the case of binary transmission, provided that the decision threshold is placed midway between the two nominal received signal voltages. For a multilevel transmitted signal comprising m levels (and so requiring $m - 1$ decision thresholds at the receiver), it can be shown (Bennett and Davey, 1965) that for a given symbol error probability an improvement in signal-to-noise ratio of approximately $(m^2 - 1)/3$ is required relative to the binary case. (The analysis assumes a constant mean signal power and that all transmit levels are equiprobable.)

The steepness of the curve in Fig. 1.4 means that small changes in signal-to-noise ratio have a significant effect on error probability. Thus digital systems need an adequate design margin to ensure both short- and long-term reliability. We also see that multilevel systems require much better signal-to-noise ratios. This may be partially compensated by the narrower bandwidth requirements made possible by a trade-off between the number of transmit levels and the symbol rate. However, for baseband systems the balance is usually in favour of binary or three-level transmission at the most.

1.4 Waveform shaping and bandwidth requirements

In order to use the transmission medium efficiently and to reduce the received noise, the channel bandwidth should be minimized. However, this conflicts with the requirement to restrict the spreading of the pulse waveform at the receiver. For reliable decision-making, a transmitted symbol must be shaped by the channel and its associated equalizer so that it does not interfere with other received symbols. Figure 1.5 shows a common overall (channel plus equalizer) pulse shaping objective. Symbols, in the form of full width pulses in this example, are transmitted as required with time spacing T, and shaped before presentation to the

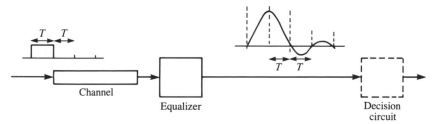

Figure 1.5. Received pulse shaping for zero intersymbol interference

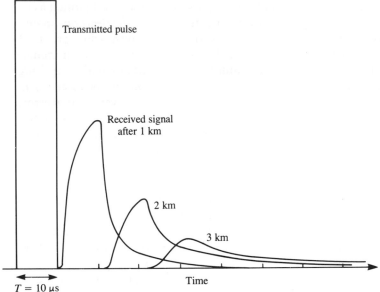

Figure 1.6   Effect of cable dispersion on pulse transmission (for 0.4 mm copper pair cable)

decision circuit to yield a peak (at the time a decision is made), with zero amplitude at all other decision times. Such a signal is said to exhibit zero intersymbol interference (i.s.i.) with respect to neighbouring received signals.

A measure of the task required of the equalizer can be seen by studying Fig. 1.6. This shows the result of transmitting a single pulse over a telephone cable representative of the copper pair type used for local line distribution. A typical pulse transmission rate for this kind of application is 100 kbit/s, that is $T = 10\,\mu$s. Pulse amplitude loss would be corrected by the amplifier of Fig. 1.3 while the equalizer would be required to compensate for the dispersion (pulse spreading) effect of the cable to the extent that subsequent pulses (arriving at 10 µs intervals in this example) did not suffer excessive i.s.i.

Nyquist (1928) showed that the minimum-bandwidth transmission characteristic (channel plus equalizer) which meets the zero i.s.i. condition is one which passes all frequencies up to $1/2T$ and stops all others. The Fourier transform of such a function yields the required impulse response:

$$\text{rect}\,fT \leftrightarrow \frac{1}{T}\,\text{sinc}\,\frac{t}{T} \qquad (1.3a)$$

This is illustrated by the frequency and time domain pair of Fig. 1.7(a).

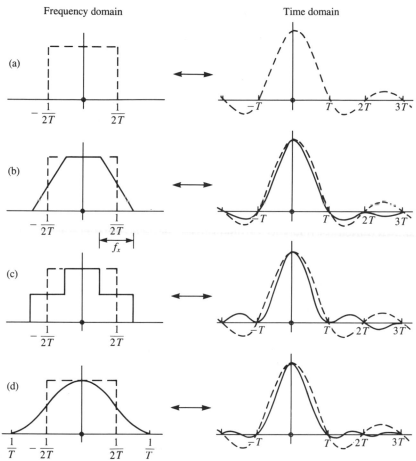

Figure 1.7 Some skew-symmetric transmission characteristics giving zero intersymbol interference: (a) Nyquist; (b) Nyquist convolved with a narrow rect function; (c) Nyquist convolved with an impulse function; (d) Nyquist convolved with a truncated cosine

That is, for a bandwidth $1/2T$, we can transmit at $1/T$ symbols per second without incurring intersymbol interference at the decision times T, $2T$, $3T$, etc. This result is only of theoretical interest because a transmission characteristic which passes all frequencies up to $1/2T$ and then exhibits an infinitely sharp cut-off cannot be realized in practice. Furthermore, small errors in decision timing can lead to large amounts of i.s.i. Both problems can be overcome, but only at the expense of extra bandwidth. A popular solution, which preserves the zero amplitude points, is to provide an equalizer such that the overall transmission characteristic (transmit pulse spectrum plus channel plus equalizer) is the result of convolving the Nyquist minimum bandwidth channel with an even frequency function. Some examples of this (skew-symmetric) class are:

$$\text{rect}\,fT * \frac{1}{f_x}\,\text{rect}\,\frac{f}{f_x} \qquad\qquad \leftrightarrow \frac{1}{T}\,\text{sinc}\,\frac{t}{T}\cdot\text{sinc}\,f_x t \qquad (1.3b)$$

$$\text{rect}\,fT * \left\{\delta\!\left(f - \frac{f_x}{2}\right) + \delta\!\left(f + \frac{f_x}{2}\right)\right\} \leftrightarrow \frac{1}{T}\,\text{sinc}\,\frac{t}{T}\cdot\cos\pi f_x t \qquad (1.3c)$$

$$\text{rect}\,fT * \left\{\frac{\pi}{2}\cos\pi fT\cdot\text{rect}\,fT\right\} \qquad \leftrightarrow \frac{1}{T}\,\text{sinc}\,\frac{t}{T}\cdot\frac{\cos\pi t/T}{1 - (2t/T)^2} \qquad (1.3d)$$

These transmission characteristics and their corresponding impulse responses are also shown in Fig. 1.7. It will be seen that all have time domain impulse responses which remain within the sinc pulse envelope (shown dotted for reference throughout). Characteristic (c) is of particular interest because it provides good tolerance to timing jitter by minimizing the mean square i.s.i. for small timing offsets (Franks, 1968). The discontinuities of (b) and (c) will be smoothed out in practical approximations. Practical equalizer design considerations often lead to adopting a raised cosine shape in the frequency domain. A full (100 per cent) raised cosine is shown as the solid lines on Fig. 1.7(d) This is seen to have a compact time response in that it always remains well within the magnitude of the Nyquist pulse envelope. Thus, at the cost of more bandwidth, tolerance to (small) timing errors is good. The full raised cosine has the added feature that the waveform at time $T/2$ is half the peak amplitude. This leads to good eye shape (see Sec. 1.5), it also has application to some partial response systems (Bennett and Davey, 1965).

It should be noted that all the characteristics of Fig. 1.7 imply linear phase properties. Uncontrolled deviations from this condition can lead to severe waveform distortion (Sunde, 1961). Equalizers must therefore take account of both the phase and the amplitude of the channel that is to be compensated. Practical solutions, which can only approximate to the theoretical requirements, aim to give acceptably low (but usually not zero) i.s.i. The design of equalizers thus involves several compromises, an issue that is dealt with in more detail in Chapter 3. We also note that it is possible to compensate (either fully or partially) for the dispersion caused by the transmission channel by appropriate shaping of the transmitted pulse before it is launched into the channel. Some line codes aim to do this, these are also discussed in Chapter 10.

1.5 The eye diagram

There are two types of degradations in a digital system:

- *Deterministic degradations* such as errors in equalizing, offsets in decision timing, amplifier gain errors and possibly transmission echoes.
- *Stochastic degradations* such as noise, interference, crosstalk and timing jitter.

Some degradations, it could be argued, should appear in both categories.

The deterministic degradations are conveniently assessed by means of an eye diagram. This is obtained, on an oscilloscope for example, by writing all possible received sequences on top of each other while triggering the oscilloscope timebase from the data clock as illustrated in Figs 1.8(a) and (b).

The eye diagram provides a great deal of information about the performance of a digital system. In the absence of noise, the width of the eye opening gives the time interval over which the received signal can be sampled without error from i.s.i. The sensitivity of the system to decision timing errors is determined by the rate of closure of the eye as the sampling time is varied. The height of the eye, at a specified decision time, determines the margin over noise.

Given distribution probabilities for the noise and timing jitter, it is then possible to determine the optimum sample time and decision threshold voltage. Figure 1.8(c) gives a diagrammatic view of these points. The diagram serves to illustrate how the *worst-case signal margin* can be very much less than the ideal design value. Noting the steepness of the error probability curve (Fig. 1.4), it can be seen how worst-case conditions can dominate performance. It follows that reliable estimates for the worst-case signal margin are crucial if the performance of a system is to be predicted accurately.

1.6 Channel capacity

This introductory chapter has addressed two principal questions: how fast can we transmit digital signals over a given communication channel and how many symbol levels can be reliably distinguished at the receiver? We have seen that the answer to the first question depends upon the channel bandwidth, while the answer to the second is determined by the signal-to-noise ratio at the receiver. The two combine to give a measure of the information capacity of the communication channel. Shannon (1948) showed that: *for a channel of bandwidth B and signal-to-noise ratio S/N, the maximum capacity is given by*

$$C_{max} = B \log_2 (1 + S/N) \text{ bit/s}$$

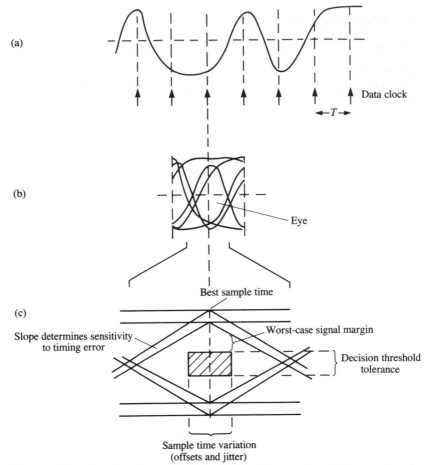

Figure 1.8 Construction of a binary eye: (a) equalized binary signal; (b) eye diagram; (c) interpretation of the eye diagram

Again, this is a maximum theoretical bound; practical systems usually achieve only a fraction of this capacity. For example, a good-quality telephone connection will provide a bandwidth of 3 kHz and a signal-to-noise ratio of 1000, that is a theoretical C_{max} of approximately 30 kbit/s. We contrast this with the 1.2 or 2.4 kbit/s data rates often provided by telephone line data modems. Higher data rates are being achieved, but only if complicated modulators (involving multilevel transmitted symbols) and sophisticated equalizers are used.

References

Beaulieu, N. C. (1989) 'A simple series for personal computer computation of the error function', *IEEE Transactions on Communications*, vol 37, no. 9, 989–91.

Bennett, W. R. and Davey, J. R. (1965) *Data Transmission*, McGraw-Hill, New York.

Carlson, A. B. (1986) *Communication Systems*, 3rd edn, McGraw-Hill, New York.

Franks, L. E. (1968) 'Further results on Nyquist's problem in pulse transmission', *IEEE Transactions on Communications*, vol. 16, 337–40.

Hamming, R. W. (1980) *Coding and Information Theory*, Prentice-Hall, Englewood Cliffs, New Jersey.

Lee, E. A. and Messerschmitt, D. G. (1988) *Digital Communication*, Kluwer Academic, Boston.

Nyquist, H. (1928) 'Certain topics in telegraph transmission theory', *Transactions AIEE*, vol. 47, 617–44.

Shannon, C. E. (1948) 'A mathematical theory of communication', *Bell System Technical Journal*, vol. 24, 379–423, 623–57. (Also 'Communication in the presence of noise', *Proceedings IRE*, vol. 27, 10–21, 1949.)

Sklar, B. (1988) *Digital Communications*, Prentice-Hall, Englewood Cliffs, New Jersey.

Sunde, E. D. (1961) 'Pulse transmission by AM, FM and PM in the presence of phase distortion', *Bell System Technical Journal*, vol. 40, 353–422.

2 The optical fibre channel

SHAMIM SIDDIQUI

2.1 Introduction

Optical fibres are now firmly established as the most versatile type of
communication channel for both short-range and long-haul
communication links, and it seems very unlikely that this situation will
change in the foreseeable future. Consequently, *optical fibre
communication systems* (OFCSs) now represent a subject of central and
all-pervading importance in modern telecommunications technology. For
this reason we devote a whole chapter in this book to providing a
concise description of the basic physical principles underlying the
operation of optical fibres as a channel for digital data transmission.

Optical fibres have acquired their present key role in
telecommunications for two main reasons, namely their truly enormous
information-carrying capacity and the low signal attenuation of the
optical fibre as a communication channel. The combined effect of these
two features is to provide very high values for the *bandwidth times
distance* (BWTD) product for optical fibre links. This, in turn, has
enabled the spacing between repeaters in an optical fibre link to be
made orders of magnitude greater than was possible in links based on
metallic conductors.

The huge potential bandwidth of OFCS comes from the high value of
carrier frequencies used in optical fibre systems. The first, second and
third generation of OFCS operate at nominal wavelengths of 0.85 μm,
1.3 μm and 1.5 μm respectively. For light of 1.5 μm wavelength the
carrier frequency is 2×10^{14} Hz. If the modulation frequency is taken
as 10 per cent of the carrier frequency then the available signal
bandwidth for the 1.5 μm wavelength OFCS is potentially 20 000 GHz.
This is clearly an enormous information-carrying capacity capable, for
example, of transmitting several million television channels. At present,
real OFCSs make use of only a minutely small fraction of this huge
bandwidth. This is because of the bandwidth limitations of present-day
optical transmitters and receivers. These limitations arise partly from the
semiconductor devices on which the transmitters and receivers are
based—light emitting diodes or semiconductor lasers for the transmitters

and PIN-diode or avalanche photodiode (APD) detectors for the receivers—but more significantly from the electronic circuitry required to drive the devices.

The low signal attenuation of currently available optical fibres represents a quite remarkable achievement by the silica glass industry. In 1968, two years after the seminal paper of Kao and Hockham (1966) which proposed the optical fibre as a communication channel, fibres drawn from then available glass had an optical attenuation in the region of 1000 dB/km. This high value was due almost entirely to absorption by impurities in the glass. A highly effective and sustained programme of research and development followed, and optical fibres for operation at 1.5 μm wavelength are now available commercially with typical attenuation values of 0.2 dB/km or less. Such low transmission losses, together with the wide bandwidths referred to above, have led to the demonstration of OFCS transmission, for example, at a signalling rate of 80 Gbit/s over a distance of 80 km in one case and at 5 Gbit/s over a 3000 km link in another (Taga *et al.*, 1992). The 15 000 Gbit.km BWTD value thus achieved is a clear illustration of the overwhelming case for optical fibres as the ideal channel for long-haul communication links.

The greatly increased interrepeater spacings, and the resulting reduction in the number of repeaters, made possible by the above high values for the BWTD product, gives the optical fibre channel another important advantage, namely that of cost, over metal conductor links since the dominant contribution to the cost of a link is that of the repeater and terminal equipment rather than the channel itself.

The small size and weight of optical fibres is yet another of its advantages. The signal carrying *core* of a *singlemode* optical fibre is usually less than 10 μm in diameter. For the *multimode* fibre the core is typically 50 μm in diameter. For both types of fibre the core is surrounded by a cladding layer with an outer diameter usually of 125 μm. With protective layers and outer sheath the overall diameter of a single strand of optical fibre is under 1 mm. The fibre core and cladding material is basically silica glass, which is a lightweight material. Given the wide signalling bandwidth of the fibre, this small size and weight represents a great advantage for the fibre over twisted wire or coaxial cable as a communication channel, especially where the amount of duct space available for the installation of new communication links is limited.

In addition to the above, optical fibres possess a number of other attractive features. They are nonmetallic, which makes them particularly suitable for use in communication links in electrically hazardous environments. The very high carrier frequencies used in an OFCS makes the link immune to interference and crosstalk in electromagnetically noisy environments. Furthermore, since the signal being transmitted over an OFCS is confined entirely within the fibre core and cladding it

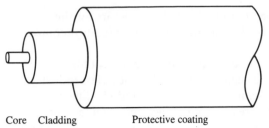

Core   Cladding   Protective coating

Figure 2.1   The basic structure of optical fibres

is not easily accessible for, possibly undesirable, interception along the length of the communication link.

2.2 Basic structure and main types of optical fibres

The basic structure of an optical fibre is shown schematically in Fig. 2.1. A circular core of refractive index n_1 is surrounded by a concentric cladding layer of refractive index n_2 with $n_2 < n_1$. The cladding layer is in turn surrounded by several protective layers usually culminating in an outer plastic sheath. The core–cladding boundary does not represent a material discontinuity. It is just the cylindrical interface where the refractive index changes radially from n_1 to n_2 in an otherwise continuous material.

The refractive index change is achieved by selectively doping the cylindrical glass preform, during its formation, from which the fibre is then drawn. At optical frequencies, a nominal value for the refractive index of pure silica glass would be 1.5.

Recall now that a coaxial metallic waveguide can have one or more electromagnetic transmission modes defined in terms of the basic transverse electric (TE) and transverse magnetic (TM) modes. Similarly, for optical fibres the size of the core and the nature of the refractive index change from n_1 to n_2 determine the three basic types of optical fibre, namely:

- Multimode step-index fibre
- Multimode graded-index fibre
- Singlemode step-index fibre

Refractive index profiles, typical core and cladding layer diameters and a schematic ray diagram representation of optical transmission in the three types of fibres is shown in Fig. 2.2.

For multimode fibres the difference between the core and cladding refractive indices is typically a few per cent. For singlemode fibres the

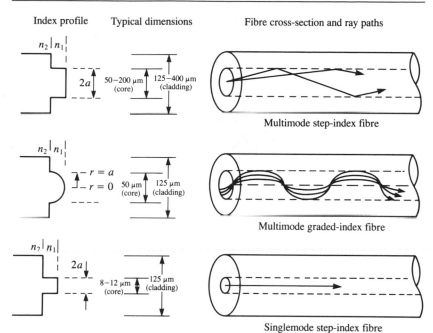

Index profile   Typical dimensions   Fibre cross-section and ray paths

Multimode step-index fibre

Multimode graded-index fibre

Singlemode step-index fibre

Figure 2.2   Refractive index profiles and typical core and cladding diameters of the three principal types of optical fibre

difference is smaller; typically between 1 and 0.2 per cent. These differences between n_1 and n_2 are frequently expressed in the form:

$$n_2 = n_1(1 - \Delta) \qquad (2.1)$$

so that for multimode fibres Δ is typically of the order of 0.02, while for singlemode fibres Δ may have a typical value of 0.005. The parameter Δ, as defined here, is called the fractional index difference. Other ways of expressing the core–cladding index difference are also used in literature.

Multimode step-index fibres are the easiest, and therefore the cheapest, of the three fibre types to manufacture. However, multimode step-index fibres suffer seriously from *intermodal dispersion* which leads to low transmission bandwidths, as we shall see. Intermodal dispersion in multimode fibres is greatly reduced by having a graded-index core as described below in Sec. 2.6. To eliminate intermodal dispersion altogether, the singlemode step-index fibre was developed. This is now firmly established as the standard telecommunications fibre. The singlemode step-index fibre too has a bandwidth limitation, provided by a different type of dispersion mechanism, namely *chromatic dispersion*. However, this is a much less severe dispersive mechanism than intermodal dispersion. The different dispersion mechanisms are described in detail later.

The optical fibre functions as a communication channel basically by

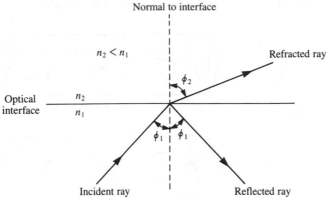

Figure 2.3 Reflection and refraction at an optical interface

the process of repeated total internal reflection of the optical frequency carrier wave at the core–cladding interface. To describe the transmission properties of the three main types of optical fibre we begin by recalling some basic definitions relating to refractive index and the phenomenon of total internal reflection.

2.3 Basic ray optics

2.3.1 Refractive index

The refractive index, n, of an optical medium is defined as:

$$n = \frac{\text{velocity of light in vacuum}}{\text{velocity of light in medium}} = \frac{c}{v}$$

Therefore:

$$v = \frac{c}{n}$$

Now $v \leqslant c$, therefore $n \geqslant 1$.

2.3.2 Snell's law

When a ray of light is incident on the interface between two optical media, the angles of incidence and refraction are related by Snell's law. Using the symbols defined in Fig. 2.3, Snell's law may be expressed as:

$$n_1 \sin \phi_1 = n_2 \sin \phi_2 \tag{2.2}$$

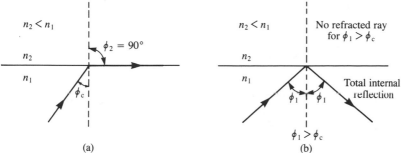

(a)   (b)

Figure 2.4   (a) For $\phi_1 = \phi_c$; $\phi_2 = 90°$. (b) Total internal reflection for
$\phi_1 > \phi_c$

2.3.3 Critical angle of incidence

If, in Fig. 2.3, $n_1 > n_2$, then there must be an angle $\phi_1 < 90°$ for which
$\phi_2 = 90°$ (Fig. 2.4a). This critical angle of incidence, ϕ_c, is given by
Snell's law as:

$$\phi_c = \arcsin \frac{n_2}{n_1} \qquad (2.3)$$

2.3.4 Total internal reflection

If in Fig. 2.3, $n_1 > n_2$ and $\phi_1 > \phi_c$, then there is no refracted ray and all
the energy in the incident ray is reflected back into the first medium
with the angle of reflection equal to the angle of incidence (Fig 2.4b).
This is known as total internal reflection. Total internal reflection is the
basis of the operation of optical fibres as waveguides.

2.4 The step-index optical fibre

We begin our description of the transmission properties of optical fibres
with the basic step-index fibre. However, before describing the
propagation properties of the step-index fibre we need to do three
things. We need to define two important more general parameters,
namely the numerical aperture (NA) and the cone of acceptance of a
fibre, which determine the ease or otherwise with which light can be
coupled into a fibre from an outside source such as an optical
transmitter. Finally, we need to introduce the concept of propagation
modes in step-index fibres.

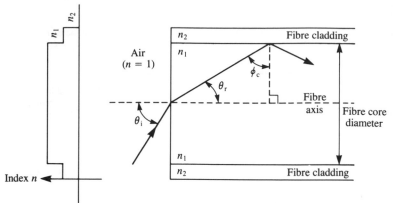

Figure 2.5 Ray diagram description of coupling light from outside into a step-index fibre. (Core and cladding outer diameters not to scale.) The refractive index profile is shown on the left

2.4.1 Numerical aperture and the cone of acceptance

Figure 2.5 shows a light ray entering the flat end of a step-index fibre at an angle of incidence θ_i such that inside the fibre it is incident on the core–cladding interface at the critical angle of incidence, ϕ_c. The angle θ_i is an important parameter for the fibre, since any light ray incident at the fibre end face at an angle of incidence greater than θ_i will then be incident on the core–cladding interface at an angle of incidence less than the critical angle and will therefore not propagate along the fibre core but will be refracted into the cladding and absorbed within the cladding and the protective layers beyond it. In other words the fibre will act as a waveguide only for light that enters the fibre from outside in a direction within an imaginary cone of apex angle equal to $2\theta_i$, with its apex on the fibre end face at the core centre. This imaginary cone is called the *cone of acceptance* of the fibre.

The value of the angle θ_i can be calculated by applying Snell's law to the propagation of the light ray in Fig. 2.5 at the air–fibre and the core–cladding interfaces. Taking the refractive index of air as unity and using the definition of the critical angle ϕ_c given by Eq. (2.3) the angle θ_i is easily shown to be given by:

$$\sin \theta_i = (n_1^2 - n_2^2)^{\frac{1}{2}}$$

$\sin \theta_i$ is referred to as the *numerical aperture* (NA) of the fibre. From the above therefore:

$$\mathrm{NA} = \sin \theta_i = (n_1^2 - n_2^2)^{\frac{1}{2}} \simeq n_1(2\Delta)^{\frac{1}{2}} \qquad (2.4)$$

where Δ is the index difference defined by Eq. (2.1). Taking n_1 to be 1.5 and Δ to be 0.02 for the multimode and 0.005 for the singlemode fibre, Eq. (2.4) gives numerical aperture values of 0.3 and 0.15 as typical for

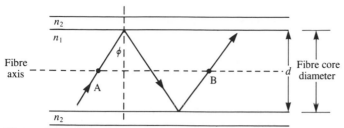

Figure 2.6   Condition for a propagating mode. Electric waves at points A and B must be in phase

the two cases. The corresponding cones of acceptance then have apex angles of 17.5° and 8.6° respectively. In order to maximize the light coupling efficiency between a light source and an optical fibre, it is clearly desirable to have as large a value as possible for the numerical aperture of the fibre. Now Eq. (2.4) indicates that the greater the difference between the core and cladding refractive indices, the greater will be the numerical aperture of the fibre. Unfortunately, this does not provide a means of achieving high numerical aperture values in practice, because a bigger difference between the core and cladding refractive indices also leads to a stronger intermodal dispersion effect which limits the transmission bandwidth of the fibre as described in Sec. 2.5.2.

2.4.2 Propagation modes in step-index fibres

For a given wavelength λ_0 (wavelength in vacuum), not all values of ϕ greater than ϕ_c are available as directions of travel for propagation modes in a step-index fibre. This is because, as has already been noted, the optical fibre functions as a waveguide by repeated total internal reflection. This leads to the possibility of destructive interference between electric fields associated with different parts of an optical ray travelling through the fibre. This would happen if the electric fields were not in phase with each other. However, for light of a given wavelength, there exists a discrete set of directions of travel for the light in the fibre for which this destructive interference does not take place. Each such direction of propagation is said to represent a propagation mode of the fibre, *at that wavelength*. The directions of travel of the propagation modes can be obtained by expressing the condition required to avoid destructive interference in the following simple geometric way. Figure 2.6 shows the path of a light ray travelling along the core of a fibre by repeated total internal reflection taking a path specified by the angle of incidence ϕ. For this path not to suffer destructive interference between electromagnetic fields from different path segments the electromagnetic waves associated with the optical ray at points A and B must be in phase, the optical path AB representing one complete round trip across the fibre core cross-section. Therefore propagating modes are defined by the condition:

$$2\pi \cdot \frac{2n_1 d}{\lambda_0 \cos \phi_m} - 2\delta = 2\pi m \tag{2.5}$$

where δ is the phase shift at each total internal reflection, m is an integer and λ_0 is the light wavelength in air. The core refractive index, n_1, appears because we are considering propagation inside the core and λ, the light wavelength inside the fibre core, is related to λ_0 by $\lambda = (\lambda_0/n_1)$.

Equation (2.5) leads to discrete values of ϕ which represent the direction of propagation of each allowed mode. We may express Eq. (2.5) as:

$$\cos \phi_m = \frac{2n_1(d/\lambda_0)}{m + (\delta/\pi)}$$

Now δ is always less than 2π for non-normal incidence, and since m will be much greater than unity then:

$$\cos \phi_m = \frac{2n_1}{m} \left(\frac{d}{\lambda_0} \right) \tag{2.6}$$

Recall from Sec. 2.6 that what was described as the multimode step-index fibre had a core diameter in the region of 50 μm. Therefore Eq. (2.6) indicates that, for light of wavelength in the 1 μm region, such a fibre has a very large number of discrete propagating modes, hence its name. Now each mode has its own direction of propagation and this leads to *modal dispersion*, which places a severe limitation on the usable transmission bandwidth of a multimode step-index fibre, as we now describe.

2.5 Optical pulse distortion in step-index fibres

Modal dispersion is the main mechanism responsible for optical pulse distortion in multimode step-index fibres. Consider a rectangular optical pulse used in an optical binary digital transmission system. When the pulse is launched into the fibre at the transmitter end, the total pulse energy is coupled into more than one propagating mode of the fibre with each mode having a different direction of travel from the others. Therefore, for the same external length of fibre, the light ray path length is different for each propagating mode, and since all modes have the same ray path velocity, this leads to the optical pulse being broadened in the time domain when it arrives at the receiver end of the fibre link.

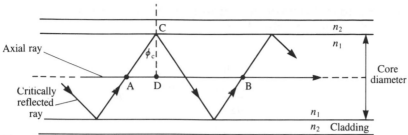

Figure 2.7 Optical paths of the axial ray and the totally internally reflected ray in a multimode step-index fibre

This is referred to as modal dispersion which we now describe quantitatively.

2.5.1 Modal dispersion

Figure 2.7 shows the paths of the axial ray and the totally internally reflected ray, at the critical angle of incidence, propagating through a multimode fibre. If we consider all the allowed propagation modes of the fibre, it is clear that the axial ray has the minimum transit time and the critically reflected ray the maximum transit time for travelling across a given external length, such as AD, of the fibre. These maximum and minimum transit times, T_{max} and T_{min}, and their difference are given by:

$$T_{max} = \frac{n_1 \cdot AC}{c} = \frac{n_1 \cdot AD}{c \sin \phi_c} = \frac{AD \cdot n_1^2}{cn_2} \quad \text{since } \sin \phi_c = \frac{n_2}{n_1}$$

$$T_{min} = \frac{n_1 AD}{c}$$

$$\Delta T = T_{max} - T_{min} = \frac{AD \cdot n_1}{c}\left(\frac{n_1}{n_2} - 1\right)$$

and:

$$\frac{\Delta T}{T_{min}} = \left(\frac{n_1}{n_2} - 1\right) = \left(\frac{n_1 - n_2}{n_2}\right) \tag{2.7}$$

Equation (2.7) is an expression of modal dispersion. The axial transit time, T_{min}, is referred to simply as *the* transit time, T, and we can use the above expressions to calculate the limitation imposed by modal dispersion on the transmission bandwidth of the step-index fibre.

2.5.2 Bandwidth limitation due to modal dispersion in step-index multimode fibres

For a fibre link of length L, the transit time, T, is equal to $(n_1 L/c)$, where c is the velocity of light in air. From Eq. (2.7) therefore:

$$\Delta T = T\left(\frac{n_1 - n_2}{n_2}\right) = \frac{L\, n_1}{c\, n_2}(n_1 - n_2) \tag{2.8}$$

Now, in order to limit intersymbol interference to an acceptable level, if we assume the maximum practicable bandwidth, B, to be:

$$B \leqslant \frac{1}{4\Delta T}$$

then Eq. (2.8) gives:

$$B = \frac{c\, n_2}{4L\, n_1}\left[\frac{1}{(n_1 - n_2)}\right] \tag{2.9}$$

For a fibre of length 1 km and index difference 1 per cent, the usable bandwidth is therefore:

$$B = 5\,\text{Mbit/s}, \qquad \text{taking}\,\frac{n_2}{n_1} \approx 1 \quad \text{and} \quad n_1 = 1.5$$

This value of 5 Mbit/s is to be compared to the potential signalling bandwidth of 20 000 GHz referred to earlier. The conclusion is that modal dispersion in a multimode step-index fibre restricts the usable bandwidth to such an extent that this type of optical fibre offers no clear advantage over metal coaxial cables as far as bandwidth is concerned.

Now Eq. (2.9) suggests that the bandwidth could be increased by reducing the index difference $(n_1 - n_2)$. However this is not a satisfactory solution to the problem, for two reasons. Firstly, recall that according to Eq. (2.4), increasing B by decreasing the index difference would also lead to a reduction in the numerical aperture of the fibre, which in turn would reduce the efficiency of coupling light into the fibre core to unacceptably low levels. Secondly, there are practical limitations on fabricating fibres with index differences much below 1 per cent.

A better approach to reducing modal dispersion in multimode fibres is to use graded-index rather than step-index fibre.

2.6 The graded-index multimode fibre

The origin of modal dispersion in step-index multimode fibre is the fact that the energy of an optical binary pulse is carried by many propagating

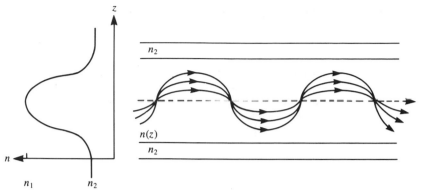

Figure 2.8   The graded-index multimode fibre

modes which all have different ray path lengths but all travel with the
same ray velocity. In graded-index multimode fibres the objective is to
arrange for a correspondingly higher ray velocity for propagating modes
with the longer ray paths and thus, ideally, to equalize the overall fibre
transit time of all the propagating modes. This is done by radially
grading the fibre core refractive index, in a highly controlled way, from
a maximum at the core centre down to the cladding index value at the
core–cladding interface as indicated in Fig. 2.8.

Now the ray velocity in any particular region of the core will be equal
to c/n where n is the refractive index in that region. Consequently, the
further away a ray path is from the fibre axis, the greater the ray
velocity. By choosing the right refractive index profile it is possible, in
principle, to reduce ΔT to zero in this way. A quantitative analysis
shows the required refractive index profile to be a shape very close to a
parabola and graded-index multimode fibres with such index profiles
have been available commercially for some time.

Although the graded-index design has greatly reduced the level of the
modal dispersion problem from that in a step-index fibre, it has not
solved the problem completely, for a combination of reasons. Firstly,
the modal dispersion parameter, ΔT, in the vicinity of $\Delta T = 0$, is a
resonantly sensitive function of the shape of the index profile curve, so
that one has to get the index profile exactly right. Furthermore, in
practice, index profiles suffer from a number of imperfections due to
various aspects of the fibre manufacturing process. Graded-index
multimode fibres therefore exhibit a small residual amount of modal
dispersion as indicated later in the summary of typical fibre parameters
(Table 2.1).

The modal dispersion problem was overcome by the development of
the singlemode step-index fibre, whose structure is such that it can
support only one propagating mode.

Table 2.1 Typical fibre parameters

		Wavelength (μm)		
Fibre Type	Parameters	0.85	1.3	1.55
Singlemode	Loss (dB/km)		0.4–0.7	0.2–0.3
	Dispersion (ps/km.nm)		1–2	15–20
Multimode	Loss (dB/km)	2–4	0.7–1.5	0.4–0.7
(Graded-index)	Dispersion (ps/km.nm)	100–200	5–6	15–20

2.7 The singlemode step-index fibre

2.7.1 Wave representation of optical propagation

The ray optics representation we have used so far is really valid only for flat slab-like waveguides. For an adequate description of optical propagation in real optical fibres it is necessary to use the electromagnetic wave representation of optical propagation. This is done by solving Maxwell's equations in cylindrical coordinates with the boundary conditions appropriate to the fibre core–cladding interface. We shall not carry out such an analysis here. Instead we shall simply quote and use the main result of such an analysis. The purpose of this analysis is to establish the condition under which a step-index fibre can sustain only one propagating mode. The analysis based on Maxwell's equations expresses this singlemode condition in terms of a fibre parameter called the normalized frequency of the fibre, usually represented by the symbol V.

2.7.2 The normalized frequency

The normalized frequency, V, also referred to as the V-number or V-parameter of a step-index fibre, is a dimensionless number defined by:

$$V = \frac{2\pi a}{\lambda} (n_1^2 - n_2^2)^{\frac{1}{2}} \tag{2.10}$$

where λ is the operating wavelength inside the fibre core and a is the core radius.

2.7.3 The single propagation mode condition

A step-index optical fibre, characterized by the core radius a, n_1 and n_2, has only one propagating mode for light of wavelength λ if

$$V = \frac{2\pi a}{\lambda} (n_1^2 - n_2^2)^{\frac{1}{2}} \leqslant 2.405 \qquad (2.11)$$

This is the single propagating mode condition for step-index fibres. If $V > 2.405$ then the fibre can sustain more than one propagating mode for light of wavelength λ. In other words, $V > 2.405$ characterizes multimode transmission while $V \leqslant 2.405$ characterizes singlemode transmission of optical signals in a step-index fibre. Derivations of Eq. (2.10), which defines the parameter V, and the singlemode condition $V \leqslant 2.405$ quoted in Eq. (2.11), may be found in any one of the textbooks listed at the end of this chapter.

2.7.4 The cut-off wavelength

Using the single propagating mode condition (2.11), the cut-off wavelength, λ_c, of a step-index fibre is defined as:

$$\lambda_c = \frac{2\pi a}{2.405} (n_1^2 - n_2^2)^{\frac{1}{2}} \qquad (2.12)$$

From this definition of the cut-off wavelength, it follows that a step-index fibre is a singlemode fibre for wavelengths greater than or equal to λ_c, and a multimode fibre for propagation of wavelengths smaller than λ_c.

2.7.5 Structure of singlemode fibres

The single propagating mode condition (2.11) can be satisfied by making either (a/λ), or $(n_1^2 - n_2^2)^{\frac{1}{2}}$, or both, sufficiently small. We have already noted the limitation on how small a value $(n_1^2 - n_2^2)^{\frac{1}{2}}$ can have in a real and practicable fibre. Taking an index difference of 0.2 per cent as a realistically small value, we have that for $n_1 = 1.53$, say:

$$(n_1^2 - n_2^2)^{\frac{1}{2}} \approx 0.08$$

Substituting this into Eq. (2.11) leads to the result that, in this case, to have a singlemode fibre we need:

$$\left(\frac{d}{\lambda_0}\right) \leqslant 6 \qquad \text{or that} \qquad d \leqslant 6\lambda_0$$

where λ_0 is the operating wavelength in air and d is the core diameter.

The first generation of optical fibre communication systems operated at approximately 0.85 μm wavelength. The next two generations operated at 1.3 μm and 1.5 μm wavelengths respectively. Singlemode fibres were not produced for $\lambda = 0.85$ μm systems for three main reasons:

1   First-generation 0.85 μm optical sources were light emitting diodes (LEDs), with a relatively broad optical spectrum, so that chromatic dispersion (see Sec. 2.8) would still have induced a considerable amount of optical pulse spreading.

2   Optical attenuation at 0.85 μm is relatively high. Therefore 0.85 μm systems could only be used for short links in any case.

3   Fabricating fibres with core diameters less than 4 μm is difficult now and was very difficult then.

Singlemode fibres were therefore produced for operation at 1.3 μm and 1.5 μm wavelengths only. These have core diameters in the range 6–10 μm, with a cladding outer diameter of 125 μm. These are now referred to as standard telecommunications fibre.

2.8 Chromatic dispersion in singlemode fibres

By definition, singlemode fibres do not suffer from modal dispersion. There is, however, another dispersion mechanism, namely chromatic dispersion, which determines the *bandwidth times distance* product for singlemode fibres. This is a much weaker dispersion mechanism than modal dispersion. Consequently, singlemode fibres have very much greater transmission bandwidths than multimode fibres. Chromatic dispersion is also present in multimode fibres but its effect may be ignored because of the much greater contribution made to optical pulse spreading by modal dispersion.

Chromatic dispersion occurs because real optical sources have nonzero spectral width, and the resulting spread in λ values gives rise to a progressive broadening of the optical pulse as it propagates down a fibre. There are two quite different kinds of chromatic dispersion, namely waveguide dispersion and material dispersion. We now consider them in turn.

2.8.1 Waveguide chromatic dispersion

Recall from Eq. (2.6) that for a step-index fibre the permitted angles of propagation, ϕ_m, are given by:

$$\cos \phi_m = \frac{2n_1}{m}\left(\frac{d}{\lambda}\right) \qquad (2.13)$$

Now if we have more than one wavelength present (λ_1, λ_2, λ_3, . . .) then each wavelength will have its own value of ϕ_m according to Eq. (2.13). Consequently, there are now as many, closely spaced, propagation directions associated with the mode integer m, as there are wavelengths. This will clearly give rise to a modal dispersion effect in the same way as for multimode fibres. Since this dispersion effect arises from a combination of the waveguide nature of the fibre and the presence of more than one wavelength in the propagating light, it is referred to as *waveguide chromatic dispersion*. This is a much smaller dispersion effect than the modal dispersion due to modes with different m-values in a multimode fibre.

In the above we have assumed that, in Eq. (2.13), n_1 has the same value for all the different wavelengths, λ_1, λ_2, λ_3, ... In other words, we have assumed no wavelength, or chromatic, dispersion in the actual optical material which forms the core of the fibre. This is, of course, not the case in a real fibre. In real fibres n_1 is a function of wavelength (i.e., n_1 is $n_1(\lambda)$) and in fact the dispersion due to the difference in the velocity of light arising from the different n_1 values associated with each wavelength gives rise to a material dispersion, which is described further in Sec. 2.8.2, and which is greater than the waveguide dispersion we are considering here.

The two important points to note about waveguide dispersion are therefore that, provided there is more than one wavelength present, then:

- Waveguide dispersion will be present even if the core medium is nondispersive, i.e., even if n_1 is independent of λ.
- If the medium is dispersive (i.e., if $n_1 = n_1(\lambda)$) then the material dispersion associated with $n_1(\lambda)$ is greater than the pure waveguide dispersion, because of the different values of λ present.

It is this material chromatic dispersion which we consider next.

2.8.2 Material chromatic dispersion

In real fibres the core refractive index n_1 is a function of λ. This means that the velocity of light in the core varies with λ. Consider an optical transmitter source with a central wavelength λ, in air, and a spectral linewidth $\Delta\lambda$ under modulation. Assume the transmitter emits an optical pulse, of negligible width in the time domain, and let this pulse be launched into a monomode fibre link of length L as indicated in Fig. 2.9. In the frequency (or wavelength) domain, the pulse energy is carried through the fibre by the range of wavelength values contained within $\Delta\lambda$, and since the different wavelengths have different velocities through the core material, arrival times of the different wavelengths at the output of the fibre will be staggered. In the time domain, therefore, the

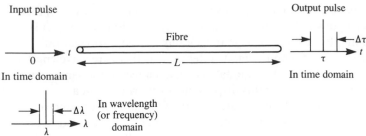

Figure 2.9 Optical pulse spreading resulting from material chromatic dispersion

output optical pulse will have acquired a spread, $\Delta\tau$, centred on τ, the time of arrival of λ, the centre wavelength. The pulse spread, $\Delta\tau$, after propagation through a fibre of length L, due to an optical source wavelength spread $\Delta\lambda$, is given by:

$$\Delta\tau = -L\left(\frac{\lambda}{c}\frac{d^2n}{d\lambda^2}\right)\Delta\lambda \qquad (2.14)$$

The term in parentheses is called the *material dispersion* of the fibre and is an important fibre parameter. Its value is usually quoted, for a given value of λ, in units of ps/nm.km. The pulse spread, $\Delta\tau$, calculated through Eq. (2.14) for a given source–fibre combination, can then be used to calculate the chromatic dispersion limited bandwidth of the link in the same way as was done for the modal dispersion case in Sec. 2.5.2. It is left as an exercise for the reader to calculate the chromatic dispersion limited bandwidths of singlemode fibre links using values for material dispersion quoted in Table 2.1 at the end of this chapter, and to compare them with the bandwidth obtained in Sec. 2.5.2 for the step-index multimode fibre. Optical transmitters with spectral widths, $\Delta\lambda$, of the order of 0.1 nm are now widely available commercially.

2.8.3 Total chromatic dispersion

The total chromatic dispersion exhibited by a fibre is the sum of the waveguide and the material chromatic dispersions described above. Figure 2.10 shows the wavelength dependence of total chromatic dispersion and its two constituent components for singlemode silica fibres. The material component goes through zero at about 1.273 µm wavelength. Note that negative values of material dispersion do not, unfortunately, mean the possibility of pulse compression rather than pulse spreading at those wavelengths. A negative dispersion value indicates only that shorter wavelengths have a higher velocity than longer ones in that wavelength region, the situation being the other way

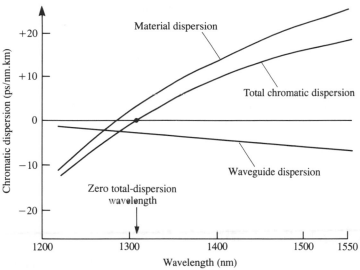

Figure 2.10   Chromatic dispersion in singlemode step-index silica fibres

round for positive dispersion values. The total chromatic dispersion for the fibre is zero at a wavelength very close to 1.3 μm. This is fortuitous since the optical attenuation of silica has a minimum at this wavelength. This led to 1.3 μm being chosen as the operating wavelength for the second generation of optical fibre communication systems. There is however another, even deeper, attenuation minimum at 1.5 μm wavelength and to exploit this low attenuation value the third-generation OFCSs operate at this wavelength despite the fact that, for the standard fibre, the total chromatic dispersion at 1.5 μm is in the region of 15–20 ps/nm . km. Consequently, in designing a transmission link using standard telecommunications fibre the first decision that needs to be made is the choice between 1.3 μm and 1.5 μm as the operating wavelength. Clearly, the decision will be determined by whether it is more important to maximize the length of the link or its signalling bandwidth. The need for this trade-off can be avoided, at a price, by the use of dispersion-shifted or dispersion-flattened fibres.

2.8.4 Dispersion-shifted and dispersion-flattened fibres

In silica fibre design it is not fruitful to attempt to modify the material chromatic dispersion contribution to the total fibre dispersion. The waveguide dispersion component, on the other hand, is amenable to considerable manipulation by departing from the simple step-index design. The first stage in this direction was to devise a refractive index profile which would simply increase the negative slope of the waveguide dispersion curve in Fig. 2.10. By achieving this in a controlled way it is now possible to produce fibres whose total chromatic dispersion is zero at any specified wavelength between 1.3 μm and 1.65 μm. Such fibres are called *dispersion-shifted fibres*. A more complicated modification of waveguide dispersion has enabled fibres to be produced that have a total chromatic dispersion of only a few ps/nm.km over the entire wavelength range between 1.3 μm and 1.5 μm. These are referred to as *dispersion-flattened fibres*. Understandably, both dispersion-shifted and dispersion-flattened fibres are considerably more expensive to produce than standard telecommunications fibre. Consequently, they have so far found application only in demonstration-type, rather than real, fibre transmission systems.

2.9 Optical attenuation in fibres

Optical loss in fibre links may be divided into *structural attenuation* and *material* or *intrinsic attenuation*. Structural attenuation is due to a range of effects. The three most important sources of structural attenuation are *microbending losses*, *coupling losses* and *splicing losses*. The first is due to propagating optical power lost into the cladding where, for example, the fibre may have been routed along a curved path and the resulting curvature at the core–cladding interface has reduced the waveguiding efficiency of the fibre. Coupling losses at fibre connectors and splicing losses where fibres are fusion joined to form long links now contribute typical values of between 0.1 and 0.2 dB per joint to the overall optical attenuation of an optical link.

2.9.1 Material attenuation

There are two types of material attenuation mechanisms, namely *material absorption*, and *scattering losses*. Material absorption is due to atomic and molecular resonant electronic absorption together with silica lattice phonon absorption. Attenuation by scattering is due to impurities and microscopic inhomogeneities in the fibre core which act as scattering centres. This is *Rayleigh scattering*, for which the scattering loss is proportional to λ^{-4}. Rayleigh scattering is responsible for the limit on the minimum optical attenuation attainable in real fibres.

A typical attenuation versus wavelength plot for standard telecommunications optical fibres is shown in Fig. 2.11. This shows the two important transmission loss minima, at 1.3 μm and 1.5 μm wavelengths. Note that these minima are very close to the Rayleigh scattering limit so that no significant further improvements can be expected over the attenuation values already achieved at these wavelengths. The λ^{-4} dependence of the Rayleigh scattering suggests

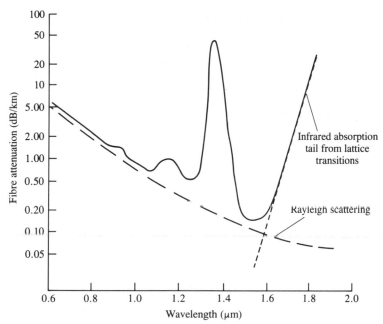

Figure 2.11 Wavelength dependence of optical attenuation in silica fibres

operation at even longer wavelengths to reduce material attenuation further. However, Fig. 2.11 also indicates that silica-based fibre cannot be used in communications systems at wavelengths greater than ~ 1.6 µm because of the steep onset of resonant lattice absorption at about 1.6 µm.

Typical transmission loss and dispersion parameters for commercially available single and multimode fibres are shown in Table 2.1 for the three principal operating wavelengths.

2.10 Summary of fibre types

- *Step-index multimode fibres*   Cheap to make, with large (~ 50–80 µm) core diameter and therefore easy to couple light into and out of. But low bandwidth due mainly to modal dispersion. Suitable for low data rate/short-distance links.
- *Graded-index multimode fibres*   Wider bandwidth in comparison with the step-index fibre, while retaining large core diameter. Easy to couple to optically. Suitable for low data rate/medium-range systems or medium data rate/short-range systems such as telecommunications junction network.
- *Step-index singlemode fibres*   Small core diameter (~ 6–10 µm), but

the problem of optically coupling to the fibre has now been solved on a commercial scale. Has wide bandwidth. Low attenuation/low chromatic dispersion trade-off between 1.3 μm and 1.5 μm wavelength fibres. Suitable for wide bandwidth/long-haul telecommunications and data links.

- *Dispersion-shifted fibres*   Small core diameter, singlemode fibre with the total dispersion zero-shifted to some specified value in the range 1.3 μm to 1.6 μm. Partial solution to the attenuation/dispersion trade-off necessary in standard singlemode fibres, compared to which it may have slightly higher attenuation and a higher price.
- *Dispersion-flattened fibres*   Singlemode fibre with only a few ps/nm . km total chromatic dispersion over the whole range between 1.3 μm and 1.6 μm. Offers really vast, wavelength-multiplexed, bandwidth over long-haul links but at a much higher price than the standard fibre.

References and further reading

More detailed accounts of optical fibres may be found in the growing number of books on optical fibre communication systems, of which the four listed below are good repesentative examples.

References

Kao, K. C. and Hockham, G. A. (1966) 'Dielectric-fibre surface waveguides for optical frequencies', *Proc. IEE*, vol. 7, 1151–8.
Taga, *et al.* (1992) Post-deadline papers PD1 and PD2 in the proceedings of the 18th European Conference on Optical Communications (ECOC '92), Berlin, 27 September–1 October 1992.

Further reading

Gowar, J. (1993) *Optical Communication Systems*, 2nd edn, Prentice-Hall, Hemel Hempstead, U.K.
Keiser, G. (1992) *Optical Fibre Communications*, 2nd edn, McGraw-Hill, New York.
Senior, J. M. (1992) *Optical Fibre Communications*, 3rd edn, Prentice-Hall, Hemel Hempstead, U.K.
Van Etten, W. and van der Platts, J. (1991) *Fundamentals of Optical Fibre Communications*, Prentice-Hall, Hemel Hempstead, U.K.

3 Equalization

LARRY LIND

3.1 Lumped equalizers

The word *equalizer* is very suggestive of what we want to accomplish, which is to make up for the transfer characteristic deficits of an existing system. For example, the modulator, amplifier, transmitter filter, channel, and receiving filter of a system will all introduce some distortion which the equalizer should correct.

In terms of frequency domain performance, the distortion comes in two forms—amplitude distortion and group delay distortion. Both are important for data transmission. The theoretically ideal Nyquist minimum bandwidth target (as introduced in Chapter 1) requires flat amplitude and group delay (or linear phase) characteristics across the passband, and infinite attenuation outside the passband. This result is seen from Fourier transform theory, which gives the Fourier pair for a low-pass channel (Carlson, 1975):

$$e^{-j2\pi f t_0} \text{ rect } (f) \Leftrightarrow \text{sinc } (t - t_0)$$

The time domain impulse response $\text{sinc}(t - t_0)$ has zero intersymbol interference (which will be discussed more fully in the next section). However, when the amplitude and group delay passband responses are only approximately flat, the system has *low* intersymbol interference, still giving good data discrimination at the receiver.

In this section two contrasting design techniques will be explored, both realizable with lumped filters. The designs are both carried out in the frequency domain.

3.1.1 'Amplitude followed by group delay' equalizers

The first design is called the classic method, because it has been around for a long time! The strategy starts by correcting the amplitude distortion, as seen in Fig. 3.1.

Put into equation form, the original amplitude response $|H(f)|$ (the verticals denote magnitude) and the amplitude equalizer $A(f)$ should ideally give:

$$|H(f)|A(f) = \text{constant} \qquad 0 \leqslant f \leqslant f_c$$

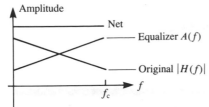

Figure 3.1 The original and equalizer amplitude characteristics combine to produce a flat response over the passband $0 \leqslant f \leqslant f_c$

Circuit theory approximation methods can be used to design a passive lumped network $E_1(f)$ whose amplitude response closely matches $A(f)$ over the passband. For example, an n-element filter with transfer function $E_1(f)$ can have its amplitude response matched to $A(f)$ on n passband points, by adjustment of its element values. If this response is not sufficiently accurate, then n can be increased. Sooner or later there are bound to be enough passband colocation points such that the error over the other points is acceptably small.

Before the advent of data transmission over channels, the amplitude correction stage was probably sufficient, and the classic method would end here. However, as we have noted above, for data transmission the group delay also needs to be reasonably flat across the passband.

Although $E_1(f)$ has equalized the amplitude response, it has created additional group delay distorton. So the next step in the classic design is to equalize the accumulated group delay distortion. All-pass network sections are usually used for this purpose. Their flat amplitude response means that they do not disturb the amplitude equalization already achieved. The pole-zero pattern for a simple all-pass 'bump' equalizer is shown in Fig. 3.2. Both poles and zeros are equidistant from the frequency axis. The poles are in the left-hand plane, to ensure stability.

The group delay (GD) that results from this equalizer is displayed in Fig. 3.3. The sharpness of the peak is controlled by the closeness of the poles and zeros to the frequency axis. By using two of these sections in cascade, the equalization of Fig. 3.4 is achieved. The bumps are added

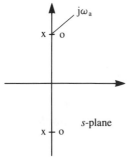

Figure 3.2 Pole-zero pattern for a second-order bump equalizer

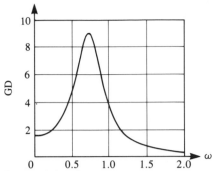

Figure 3.3 Group delay response of an all-pass bump equalizer section

to the original distorted group delay to produce a nearly flat group delay characteristic across the passband $0 \leqslant \omega \leqslant 1$.

This two-stage amplitude followed by group delay design method is safe, conservative, and extensively used in practice. The designer knows that if the current design does not quite meet the specification, more elements can be added to either network. Also, the group delay can be modified without changing the amplitude response.

We illustrate this approach with a simple numerical example. Let a low-pass channel, with a cut-off frequency of 1 rad/s, be modelled by:

$$H(s) = \frac{0.5}{s + 0.5}$$

The channel has no zeros, and one pole at $s = -0.5$. Its amplitude and group delay responses are shown in Fig. 3.5. Obviously, some equalization is needed!

The first step is to equalize the amplitude over the passband. In this

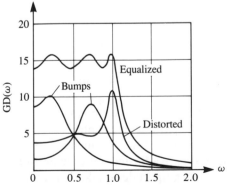

Figure 3.4 The addition of bump sections produces overall group delay equalization

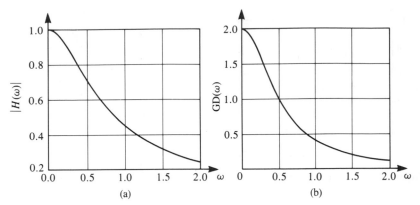

Figure 3.5 (a) Amplitude and (b) group delay responses for the low-pass channel whose cut-off frequency is required to be 1 rad/s

simple example an all-pole four pole filter will be used for correction. The all-pole filter is attractive, for it can be realized as a ladder network. To achieve the requisite design, the four poles are adjusted manually until the combined response has, for example, an equiripple passband. The strategy is as follows. If a complex conjugate pole pair is positioned at $(r \pm jx)$, r controls how much the characteristic is lifted, with the peak uplift at frequency x. The smaller r, the greater the uplift. Referring to Fig. 3.5(a), the poles are initially placed at $(-0.3 \pm j0.5)$ and $(-0.2 \pm j1.0)$. Small movements in these values are then made. After each adjustment the amplitude response is replotted, to see the effect of the adjustment. Equiripple convergence to three-place accuracy takes about 40 iterations, and gives the pole locations:

$$p_{1,2} = -0.310 \pm j0.565$$
$$p_{3,4} = -0.108 \pm j1.000$$

The resulting amplitude and group delay responses are displayed in Fig. 3.6.

Although the amplitude response is much improved, look at the group delay distortion! It is worse than the original channel group delay, having a large peak near the cut-off frequency. Group delay correction is clearly necessary. To keep the example simple, two bump equalizer sections are used. Each section consists of two poles at $(-r \pm jx)$, and two zeros at $(r \pm jx)$. The parameter r controls how much the group delay is lifted, with the peak uplift at frequency x. The smaller r, the greater the uplift. The poles are adjusted until the overall group delay response (channel + all-pole amplitude equalizer + group delay equalizer) is, say, equiripple over the passband. The poles and zeros of the bump sections are then:

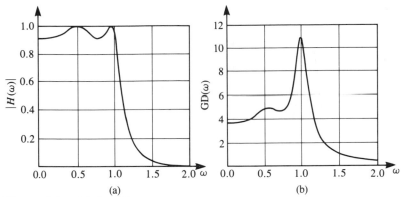

Figure 3.6 (a) Amplitude and (b) group delay response of the channel plus all-pole amplitude equalizer

$$z_{1,2} = 0.242 \pm j0.230$$
$$p_{1,2} = -0.242 \pm j0.230$$
$$z_{3,4} = 0.227 \pm j0.725$$
$$p_{3,4} = -0.227 \pm j0.725$$

The resulting overall group delay is shown in Fig. 3.7 (with the amplitude response of the network remaining the same as in Fig. 3.6a). Notice how high the mean level of group delay is now, about eight times as high as in the original channel.

Now group delay in the frequency domain is nearly the same as waveform delay in the time domain. To see this, imagine that an input signal is bandlimited to the passband of an ideal filter; then there is no amplitude attenuation of any frequency component by the filter. Also, let the filter have a phase shift equal to $e^{-j\omega t_0}$. The filter has a group delay of:

$$GD(\omega) = -d\phi(\omega)/d\omega = t_0$$

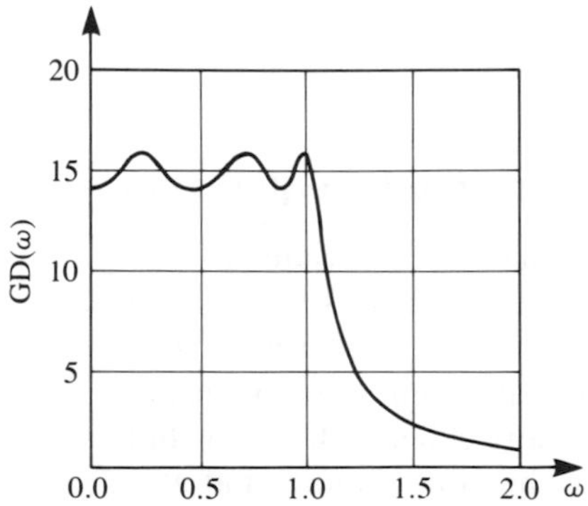

Figure 3.7 Group delay response of the overall network

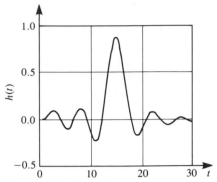

Figure 3.8 Impulse response of the overall network

By the Fourier transform shifting theorem, multiplication by $e^{-j\omega t_0}$ in the frequency domain is equivalent to shifting a waveform to the right by t_0 in the time domain, which proves the assertion.

For our example, the impulse response of the overall network is shown in Fig. 3.8. The peak of the impulse response occurs at 15 s, which is the mean level of the passband group delay in Fig. 3.7. Also, we see that the zero crossings in the response are approximately spaced by $T = \pi$ seconds (T is further explained in Chapter 1). This spacing occurs because the stopband edge has been set to $\omega_c = 1$ radian/s, or $f_c = 1/(2\pi) = 1/(2T)$ Hz.

The impulse response has low intersymbol interference. This is true because the amplitude response (Figure 3.6a) is close to a rectangular response, and the group delay is approximately flat across the passband (Fig. 3.7).

The example also shows that the all-pole amplitude equalizer creates a large group delay bump near the passband edge (Fig. 3.6b). The delay can only be equalized by bringing the rest of the passband group delay up to this peak level, creating a large amount of passband group delay. This delay can be embarrassing if the filter is part of a control loop, or is involved in two-way communication.

3.1.2 'Group delay followed by amplitude' equalizers

This large passband group delay problem can be alleviated by using a different design concept (Rhodes, 1976). Instead of using the all-pole filter as an amplitude equalizer, we use it instead as a *group delay equalizer*; then an even numerator polynomial will be employed to equalize the amplitude. The result is a more efficient design, with less time delay. The design is illustrated for the same channel that was equalized in Sec. 3.1.1:

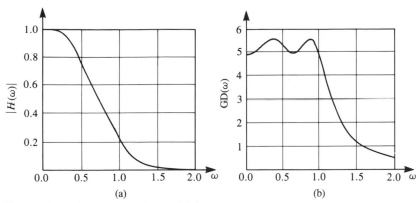

Figure 3.9 (a) Amplitude and (b) group delay responses of the channel plus all-pole group delay equalizer

$$H(s) = \frac{0.5}{s + 0.5}$$

The first step is to equalize the channel group delay. We will use a four pole all-pole filter for this purpose. Manual optimization for an equiripple response (to three-place accuracy) gives the pole locations as:

$$p_{1,2} = -0.330 \pm j0.41$$
$$p_{3,4} = -0.255 \pm j0.92$$

The net amplitude and group delay responses are shown in Fig. 3.9.

The second step is to find an even numerator polynomial to provide the amplitude correction. An even polynomial is used in anticipation of a passive filter realization. This polynomial will not contribute to group delay, and so the overall group delay will remain, as shown in Fig. 3.9(b). We use the even fourth-degree polynomial:

$$N(s) = s^4 + a_2 s^2 + a_4$$

having a quartet of zeros at ($\pm r \pm jx$). The parameter r controls how much the amplitude response is attenuated, with the peak attenuation at frequency x. The smaller r, the greater the attenuation.

Manual adjustment with two-place accuracy for, say, an equiripple amplitude response gives the zeros:

$$z_{1,2,3,4} = \pm 0.55 \pm j0.53$$

Then the overall system has the amplitude response and impulse response as shown in Fig. 3.10.

Note that now both the group delay (Fig. 3.9b) and amplitude (Fig. 3.10a) have been equalized, which is the main objective of the design. Also note that the impulse response has much less delay than previously (compare Fig. 3.10b to Fig. 3.8).

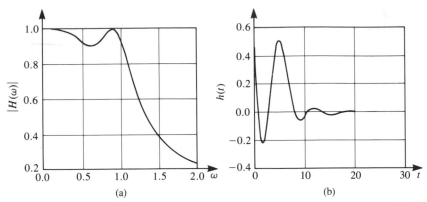

Figure 3.10 (a) Amplitude response and (b) impulse response of the overall system, using the new design method

The present impulse response contains more precursor intersymbol interference than before, but then we have not used i.s.i. as a design criterion. Indeed, we have not attempted to perform any stopband shaping at all. Normally the transmit and receive filters will provide the required stopband attenuation. Such filters have not been used in this example, for the sake of simple exposition.

3.1.3 Comparison of the two designs

The comparison is interesting. Both designs give about the same ripple levels in the amplitude and group delay passband responses. The classic method, however, yields an equalizer with eight poles and four zeros, and a time delay of 15 s. The new approach has only four poles and four zeros, and a time delay of 6 s. From an equalization point of view, the new method is more efficient, and more than halves the delay time. There are many applications where both features could be important.

In conclusion, it seems to be best to equalize first group delay with poles, then amplitude with zeros.

3.2 Discrete equalizers

There is an increasing tendency for signal processing to become digital, because of the many benefits of digital technology. So we next turn to a study of how discrete equalizers can be designed, having as an input a stream of samples of a continuous waveform. Because filter coefficients can be changed easily, these filters can be made adaptive, whereas the inductors and capacitors of analogue filters are more difficult to alter quickly. Digital designs can be realized by using a digital signal processor, which is now commonplace and inexpensive. Interestingly,

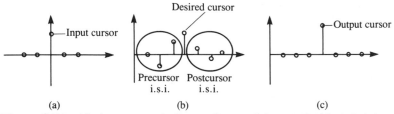

Figure 3.11 Various sampled waveforms: (a) sampled input data pulse, (b) the equalizer input, as a result of system distortion, (c) ideal equalizer output

the time domain rather than the frequency domain is used for performing these designs.

3.2.1 Introduction

A practical digital transmission system has many components that will distort the passage of a sampled pulse. There may be transmission and reception filters, modulators and demodulators, and the transmission path itself, all of which can cause distortion.

Figure 3.11(a) shows the ideal sampled input waveform for a data symbol. Each sample is shown by a small circle. A new sample occurs every T seconds. The sample sequence is nonzero at only one point, the ideal cursor. The cursor will be quantized to one of m possible levels. In this chapter only the bipolar values of ± 1 (i.e., binary transmission) will be considered, for simplicity.

Each of the sample points is used to send one bit of data. We will assume that the system is linear. Then, given the input/output performance of a single pulse, superposition can be used to find the combined effect of all the input data pulses.

The effect of system distortion (presented to the equalizer input) is shown in Fig. 3.11(b). The input pulse has been smeared out in time. There are some early nonzero samples, which represent *precursor intersymbol interference* (i.s.i.), then the *desired cursor sample point*, followed by yet more nonzero points containing *postcursor i.s.i.*

This distorted signal is fed into the equalizer. A perfect equalizer will get rid of all the precursor and postcursor i.s.i., leaving only a delayed (and possibly scaled) version of the original input cursor, as shown in Fig. 3.11(c). If a sequence of data symbol waveforms is then input, the ideal equalizer would, by superposition, produce a delayed and scaled replica of the input sequence. This output is then easily decoded by a threshold detector to reproduce the original data.

To get an idea of how to design the equalizer, we are going to make some unrealistic assumptions. At first it will be assumed that there is a strong signal, and so noise can be ignored. It will also be assumed that

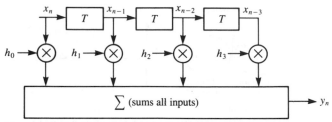

Figure 3.12   Tapped delay line equalizer structure

the input signal consists of widely spaced pulses, so that the output transient response of one pulse dies away before the next pulse appears. These assumptions will be relaxed when nonlinear equalizers are discussed.

The following sections will look at fixed and adaptive linear equalizer designs. We finish by considering a simple nonlinear design that does everything possible to achieve perfection.

3.2.2 Tapped delay line equalizer structure and operation

The four-tap structure shown in Fig. 3.12 will be used to illustrate the principles. The input x_n is a digitized sample sequence. The equalizer contains a three-section digital delay line. This could be made by using a number of three-bit shift registers, the number being equal to the word length of each input sample. Each section has a delay of T seconds. The tap outputs are fed to digital multipliers. These multipliers also have inputs h, which are to be found using the methods below. The multiplier outputs are summed together to produce the final output y_n. This structure produces a finite impulse response to the impulse of Fig. 3.11(a) and so is sometimes called an FIR filter.

Four taps are used here for simplicity. In reality, however, there can be 50 or 100 taps in some practical designs. The performance of this equalizer is determined by the tap weights, and is given by the difference equation (Oppenheim and Schafer, 1975):

$$y_n = h_0 x_n + h_1 x_{n-1} + h_2 x_{n-2} + h_3 x_{n-3} \tag{3.1}$$

If the h and x values are considered as sequences, Eq. (3.1) represents the discrete convolution:

$$y = x \otimes h \tag{3.2}$$

Generally perfection cannot be achieved with a finite length equalizer— there will always be some residual i.s.i. Even if we could design and build an infinite length equalizer (and thus achieve perfection), the argument goes that the input signal would never leave the equalizer, and so would not be of much use! However, the residual i.s.i. can be reduced to low levels with a finite length design. We will now look at two design strategies to achieve this objective.

3.2.3 The zero forcing method

With this strategy a single y output is set to one (the output cursor point), while several adjacent sample values are set to zero. That is, the output is forced to look ideal over a limited range of output values. What happens outside this range is beyond our control. It turns out, however, that this strategy does minimize the peak error of the nonzero i.s.i. points (Lucky *et al.*, 1968).

This design process is now illustrated with a numerical example. The input x is put into vector form, in anticipation of the matrix equations that follow. Let x be riddled with precursor and postcursor i.s.i.:

$$x =$$
$$[0.1250 \quad 0.2000 \quad 1.0000 \quad 0.3333 \quad 0.2500 \quad 0.1250 \quad 0.0625 \quad 0.03125]^t \quad (3.3)$$

where t indicates transpose. Thus x is a 8×1 column vector. In the development below, all such vectors will be assumed to be column vectors. The first element of x occurs at discrete time $n = 0$; $x_0 = 0.125$. The next element, $x_1 = 0.200$, occurs at time point T seconds, and so on (see Fig. 3.11). x_8 to x_∞ are assumed to be zero. We want to equalize this sequence with the four-tap equalizer of Fig. 3.12.

Although there is a general strategy, there are still some unknowns. Where should the cursor be put? What adjacent precursor and/or postcursor points should be zeroed? The best selection will be found by trial and error for this example, using a squared error criterion that is detailed later.

As there are four degrees of freedom (the h coefficients), Eq. (3.1) can be written for four successive values of n, giving four linear simultaneous equations to be solved. We start by assuming the cursor is at $y_3 = 1$, and $y_2 = y_4 = y_5 = 0$. The equation set is, in matrix form:

$$
\begin{bmatrix} y_2 \\ y_3 \\ y_4 \\ y_5 \end{bmatrix}
=
\begin{bmatrix} 0 \\ 1 \\ 0 \\ 0 \end{bmatrix}
=
\begin{bmatrix} x_2 & x_1 & x_0 & 0 \\ x_3 & x_2 & x_1 & x_0 \\ x_4 & x_3 & x_2 & x_1 \\ x_5 & x_4 & x_3 & x_2 \end{bmatrix}
\begin{bmatrix} h_0 \\ h_1 \\ h_2 \\ h_3 \end{bmatrix}
\qquad (3.4)
$$

The X matrix on the right side is a Toeplitz matrix. It has the property that all diagonals parallel to (and including) the main diagonal have a repeated element. This property leads to a simple and rapid solution algorithm for h (Press *et al.*, 1986).

Matrix inversion and multiplication gives:

Precursor zeros	1	2	3	4	5
3				84	0.17
2			550	0.027	520
1		1500	0.0121	130	620
0	5000	0.050	0.035	120	
Cursor at	1	2	3	4	5

Table contents = E_z = squared i.s.i. error in y

Figure 3.13 The total squared errors E_z as a function of the cursor location and the numbers of precursor zeros

$$h = [-0.1912 \quad 1.1442 \quad -0.3013 \quad -0.1617]^t$$

with a total squared error in y of $E_z = 0.0121$. The error includes both precursor and postcursor i.s.i., which is outside the 'ideal output' range of the zero forcing.

Since the input x has a squared i.s.i. error of 0.2497, this simple equalizer has performed well. It has reduced the squared error by a factor of 20.

Have the best cursor and zero i.s.i. points been selected? Various numerical experiments yield Fig. 3.13. This shows that the best result is the one already given (it was actually chosen after constructing the table). The output sequence for this design is calculated from Eq. (3.1) as:

$$y = [-0.0239 \quad 0.1048 \quad 0 \quad 1 \quad 0 \quad 0 \quad 0.0019$$
$$-0.0126 \quad -0.0033 \quad -0.0195 \quad -0.0051]^t$$

To achieve the best result, some precursor i.s.i. elimination is important.

The zero forcing method has an interesting interpretation in the z-plane (Oppenheim and Schafer, 1975). Let the z-transform of the input and the equalizer transfer function be $X(z)$ and $H(z)$, where $z = e^{sT}$. For the above example:

$$X(z) = 0.1250 + 0.02000z^{-1} + 1.000z^{-2} + 0.3333z^{-3} + 0.02500z^{-4}$$
$$+ 0.1250z^{-5} + 0.0625z^{-6} + 0.03125z^{-7}$$
$$H(z) = -0.1912 + 1.1442z^{-1} - 0.3013z^{-2} - 0.1617z^{-3}$$

Their product should ideally create a single nonzero output at time point nT, as seen in Fig. 3.11(c):

$$Y(z) = X(z)H(z) = kz^{-n}$$

This response has a constant magnitude and linear phase on the digital frequency axis (the z-plane unit circle $e^{j\omega t}$). In this context H is sometimes referred to as an inverse filter, as it should undo magnitude and linear phase distortion created by $X(z)$.

A plot of the roots of X (the $\times$ points) and H (the $\bigcirc$ points) in the z-plane are shown in Fig. 3.14.

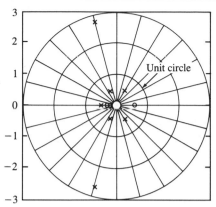

Figure 3.14 Roots of the distorted input waveform (×) and roots of the equalizer (○) combine to approximate an all-pass network

One way to achieve a constant magnitude on the unit circle is to have an infinity of zeros on a circle of nonunity radius. This can be seen from Fig. 3.14, where the h root at 0.5664 is filling in between the × s, in attempting to make a combined circle of approximately equally spaced roots. Also, the h root at -0.2629 is giving some more strength to the × root -0.4616, again to help produce a constant attenuation on the unit circle. The final h root, at 5.6813, is off the diagram and makes no obvious contribution to the constant attenuation requirement!

3.2.4 A simple adaptive strategy

Usually the equalizer algorithm in a piece of equipment is not powerful enough to solve matrix Eq. (3.4), bearing in mind the number of taps could be 100 or so. Iterative methods need to be employed. We start by assuming that h is close to its best values. How can the tap weights be adjusted to move towards the zero forced solution?

The Jacobian matrix shows the sensitivity of the sample error to changes in tap weights, by computing a matrix of first derivatives. At the solution point of Eq. (3.4), the Jacobian for the above example is:

$$
\begin{bmatrix}
\partial y_2/\partial h_0 & \partial y_2/\partial h_1 & \partial y_2/\partial h_2 & \partial y_2/\partial h_3 \\
\partial y_3/\partial h_0 & \partial y_3/\partial h_1 & \partial y_3/\partial h_2 & \partial y_3/\partial h_3 \\
\partial y_4/\partial h_0 & \partial y_4/\partial h_1 & \partial y_4/\partial h_2 & \partial y_4/\partial h_3 \\
\partial y_5/\partial h_0 & \partial y_5/\partial h_1 & \partial y_5/\partial h_2 & \partial y_5/\partial h_3
\end{bmatrix}
=
\begin{bmatrix}
0.19 & 0.23 & 0.04 & 0.00 \\
0.06 & 1.14 & 0.06 & 0.02 \\
0.05 & 0.38 & 0.30 & 0.03 \\
0.02 & 0.29 & 0.10 & 0.16
\end{bmatrix}
$$

Only y_2 to y_5 are used, because these are the outputs which are forced to look ideal by the h coefficients. The first column shows the effect on the outputs from changing h_0, the next of changing h_1, and so on. We observe that h_0 has greatest effect on y_2, h_1 on y_3, h_2 on y_4, and h_3 on y_5. This leads to the simplifying assumption that output error at a point can be lowered by changing the corresponding tap weight, and that this change will have little effect on the other points. This feature is true in general for FIR equalizers, and is sometimes referred to as the tap dominance effect (Lucky *et al.*, 1968).

Usually the Jacobian matrix is not known, but there is sufficient information (the sign of the error) to know the *direction* of change. A cautious strategy is to measure the error at each desired output point, y_2 to y_5, and then change each coefficient by a single quantum q so as to reduce the output error.

To illustrate, let the equalizer have the tap weights:

$$h = [-0.200 \quad 1.100 \quad -0.300 \quad -0.100]^t$$

and let the adjustment quantum $q = 0.001$. Applying the input sequence x to the equalizer gives the outputs:

$$y_2 = -0.0175, \quad y_3 = 0.9608, \quad y_4 = -0.0033, \quad y_5 = 0.0500$$

Using the sign of the errors as a guide (and remembering that y_3 should equal 1.0), we update h to:

$$h = [-0.200 + q \quad 1.100 + q \quad -0.300 + q \quad -0.100 - q]^t$$

The equalizer's output is then:

$$y_2 = -0.0162, \quad y_3 = 0.9622, \quad y_4 = -0.0020, \quad y_5 = 0.0497$$

Some progress has been made. All output points have been nudged closer to their ideal values. One can imagine that, even with poor starting values, after hundreds of such adjustments the equalizer will slowly but surely reach its optimal state.

This iterative algorithm does have some important advantages. It is easy to compute. It usually converges to the optimal setting (but can get trapped in a local minimum if the starting i.s.i. is unusually large (Lucky *et al.*, 1968). The main difficulty is the slow convergence rate. It is easy to think of ways of speeding up this process. For example, one could use a time-varying quantum step q, which is large at first, but then slowly shrinks to its final level.

Rather than using the fixed quantum iterative approach, a more ambitious method can be contemplated, using the above experience. The steps are as follows:

1   Observe the new output errors e_i

2   Change all tap weights by a small amount ε_i.

3   Observe the new output errors $\underline{e}_i$.

4 Assuming tap dominance (Lucky *et al.*, 1968), the sensitivity of each
output to tap movement is given by the Jacobian:

$$J_i = (\underline{e}_i - e_i)/\varepsilon_i$$

To bring $\underline{e}_i$ to 0, we use:

$$J_i = (e_i - 0)/\Delta h_i \qquad \text{or} \qquad \Delta h_i = e_i/J_i$$

Since there is some interaction, this procedure can be iterated until all
errors are acceptably small. The method uses tap dominance to
decompose a large number of simultaneous linear equations to an
equally large number of simple first-order equations.

 Although the zero forcing technique is easy to understand and apply,
it does suffer from a fundamental problem. No account is taken of the
i.s.i. that exists outside the range of perfect response. The next technique
will take *all* of the i.s.i. into account.

3.2.5 Minimum squared error

The criterion is to find the *h* coefficients which minimize the sum of all
squared precursor and postcursor i.s.i. error values, with the condition
that the cursor remains at unit height. The squared error is used for
several reasons. It leads to a simple mathematical solution. The error
surface is convex, which guarantees that an iterative solution will always
progress to the global minimum. Finally, a squared error criterion gives
greater importance to large error points, and so tends to keep the peak
errors at low values. In the following discussion, the same *x* input
sequence will be used as before to illustrate the process.

 The output *y* sequence in Eq. (3.1) can be put into the matrix form:

$$y = Xh$$

where *y* is an *n*-element column vector containing *all* nonzero outputs, *X*
is an $n \times 4$ rectangular matrix containing the nonzero sample values, *h*
is a four-element column vector (assuming the four-tap equalizer is used
again), and *n* is the number of nonzero elements in *y*. In the above
example $n = 8 + 4 - 1 = 11$.

 We start by forming E_m, the sum of the squares of all *y* values:

$$E_m = y^t y = (h^t X^t) X h = h^t (X^t X) h \qquad (3.5)$$

E_m therefore includes the main sample point. The middle portion of the
right-hand side of Eq. (3.5), $(X^t X)$, represents a 4×4 correlation matrix
R for the input *x* values. The correlation matrix has element values:

$$R_{ij} = r_i^t r_j \qquad (3.6)$$

where:

$$r_0 = [x_0 \quad x_1 \quad \dots \quad x_7]^t$$

$$r_1 = [0 \quad x_0 \quad \ldots \quad x_6]^t$$
$$r_2 = [0 \quad 0 \quad \ldots \quad x_5]^t$$
$$r_3 = [0 \quad 0 \quad \ldots \quad x_4]^t$$

If E_m were to be directly minimized with respect to h, we would get the embarrassing result $h = 0$! To avoid this trivial solution, the constraint $y_3 = 1$ is introduced (later other candidates for the cursor point will be considered) by forming the constraint equation:

$$E_c = y_3 - 1 = [x_3 \quad x_2 \quad x_1 \quad x_0][h_0 \quad h_1 \quad h_2 \quad h_3]^t - 1 = 0$$

Next, E_m and E_c are used to form an unconstrained variable E_u, by introducing a Lagrange multiplier L (Fryer and Greenman, 1987):

$$E_u = E_m - LE_c \tag{3.7}$$

The Lagrange multiplier method produces an error minimum by solving the equations:

$$\partial E_u/\partial h_0 = 0, \quad \partial E_u/\partial h_1 = 0, \quad \ldots, \quad \partial E_u/\partial h_3 = 0 \tag{3.8}$$

and by:

$$\partial E_u/\partial L = 0 \tag{3.9}$$

Some algebraic work shows that Eqs (3.8) and (3.9) lead to the matrix equation:

$$Rh = (L/2)[x_3 \quad x_2 \quad x_1 \quad x_0]^t \tag{3.10}$$

Eq. (3.10) is a set of four simultaneous linear equations, containing the unknown column vector h of equalizer coefficients. Since the Lagrange multiplier L scales all h values in Eq. (3.10) by the same amount, it can at present be set to some convenient value (say $L = 2$), and later scaled so that the desired cursor point has unity value. The R matrix is Toeplitz, which leads to a rapid solution algorithm for h (Press *et al.*, 1986).

We solve Eq. (3.10) for the input sequence:

$$x = [1/8 \quad 1/5 \quad 1 \quad 1/3 \quad 1/4 \quad 1/8 \quad 1/16 \quad 1/32]^t$$

and scale h so that $y_3 = 1$. Then the squared i.s.i. error is reduced to $E_m - y_3^2 = 0.0119$ (compared to 0.0121 for the zero forcing case). The corresponding h and y are:

$$h^t = [-0.2072 \quad 1.1481 \quad -0.2949 \quad -0.1604]$$
$$y^t = [-0.0259 \quad 0.1021 \quad -0.0145 \quad 1 \quad 0.0039 \quad 0.0024 \quad 0.0034$$
$$-0.0117 \quad -0.0026 \quad -0.0192 \quad -0.0050]$$

The error in this case is smaller than that for the zero forcing method, but only gives a slight improvement. The difference however is that the

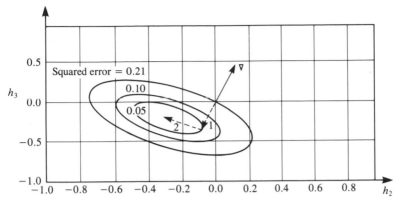

Figure 3.15 Contour plot of the squared error of the example, as a function of h_2 and h_3

present method uses *all* input samples in its computation, whereas the zero forcing method uses a limited range of values.

y_3 has been assumed to be the best candidate for the cursor point. Is this true? Numerical experiments give the following total squared i.s.i. error, as a function of which point is chosen for the cursor:

$$n = \quad 2 \qquad 3 \qquad 4 \qquad 5$$
$$E_2 - 1 = 0.0497 \quad 0.0119 \quad 0.0245 \quad 0.1663$$

So there is a best cursor point, and for this case it is y_3.

The exact solution of the minimum squared error problem requires the solution of a set of simultaneous linear equations, just as was the case for the zero forcing method.

Because of the high mathematical workload with minimum squared error (tap dominance no longer applies), a gradient solution is next considered, which solves this problem in an iterative fashion.

3.2.6 Least mean squares

(See Haykin, 1986.) We start by finding the gradient of the error function $E_m - 1$ in Eq. (3.5). Some algebraic work shows that the gradient vector is:

$$\nabla = [\partial E_m/h_0 \quad \partial E_m/h_1 \quad \partial E_m/h_2 \quad \partial E_m/h_3]^t = 2Rh \qquad (3.11)$$

This vector gives the direction of greatest increase in E_m. We are usually interested in the direction, not the magnitude of the gradient vector.

A contour plot of the squared error is shown in Fig. 3.15. The tap weights h_0 and h_1 have been fixed at their optimal values, while h_2 and h_3 have been allowed to vary. Assuming a start point at $h_2 = 0, h_3 = 0$, the figure shows the gradient ∇ at this point. To minimize the error, the best plan is to move in a direction opposite to the gradient, as shown by path

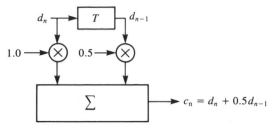

Figure 3.16   Two-path channel model

1. We would like to move along this path until the error is minimized. The process is repeated on path 2, and so on, until we arrive at the error minimum. Usually, however, the contour error plot is not at our disposal. Theoretical (Haykin, 1986) and/or experimental work can be performed, to determine a safe movement distance along the negative direction of the gradient vector. This distance is chosen to: (a) guarantee convergence, and (b) minimize the number of iterations needed to converge.

The gradient calculation in Eq. (3.11) requires the autocorrelation matrix R. This can take significant computation time, as Eq. (3.6) shows. Thus a crude approximation to R is made, by using:

$$R_a \approx xx^t \tag{3.12}$$

where:

$$x = [x(k) \quad x(k-1) \quad \ldots \quad x(k-n)]^t \tag{3.13}$$

Finally, h is updated by the equation:

$$h_{new} = h_{old} - k\nabla = h_{old} - kR_a h_{old} \tag{3.14}$$

where k, the movement distance scale factor, is chosen so as to achieve convergence in the minimum number of iterations.

This is the method of least mean squares. Although the approximation is rough, the method works surprisingly well (Haykin, 1986), and is the basis of the vast majority of equalizers in use today.

3.2.7 A non-linear equalizer

(See Cowan, 1992.) We illustrate the general concepts again with a very simple example. The input data stream d_n is made bipolar, equal to ± 1. The dispersive channel is modelled by the two-tap FIR response shown in Fig. 3.16.

This is the sort of response that could come from a two-path radio link. The main path gives an output of 1. The secondary path, representing perhaps a reflection, takes one delay longer to traverse, and has a weaker output of 0.5.

For simplicity, the equalizer memory will be limited to a single delay stage, whose output will be defined as c_{n-1}. Thus the equalizer algorithm

Table 3.1 State transition and output table for the example of Fig. 3.16

							Labels	
d_n	d_{n-1}	d_{n-2}	d_{n-3}	c_n	c_{n-1}	c_{n-2}	$c_n c_{n-1}$	$c_{n-1} c_{n-2}$
-1	-1	-1	-1	-1.5	-1.5	-1.5	D	D
-1	-1	-1	1	-1.5	-1.5	-0.5	D	C
-1	-1	1	-1	-1.5	-0.5	0.5	C	B
-1	-1	1	1	-1.5	-0.5	1.5	C	A
-1	1	-1	-1	-0.5	0.5	1.5	B	H
-1	1	-1	1	-0.5	0.5	-0.5	B	G
$-1.$	1	1	-1	-0.5	1.5	0.5	A	F
-1	1	1	1	-0.5	1.5	1.5	A	E
1	-1	-1	-1	0.5	-1.5	-1.5	H	D
1	-1	-1	1	0.5	-1.5	-0.5	H	C
1	-1	1	-1	0.5	-0.5	0.5	G	B
1	-1	1	1	0.5	-0.5	1.5	G	A
1	1	-1	-1	1.5	0.5	-1.5	F	H
1	1	-1	1	1.5	0.5	-0.5	F	G
1	1	1	-1	1.5	1.5	0.5	E	F
1	1	1	1	1.5	1.5	1.5	E	E

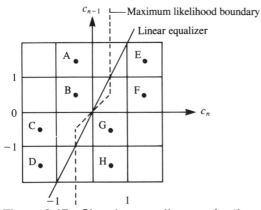

Figure 3.17 Signal space diagram for the example

will have, at time n, the inputs c_n and c_{n-1} to play with. Armed with this information, Table 3.1 can be formed. It shows the operation of the channel and the equalizer memory to various input patterns of length four. The reason for using such long sequences will become apparent later.

In forming this table, the c_{n-2} column has been formed from:

$$c_{n-2} = d_{n-2} + 0.5 d_{n-3}$$

The table shows that the channel can only produce outputs from the set

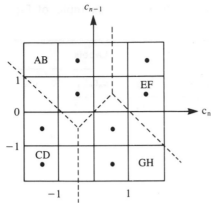

Figure 3.18   Four-region signal space

of values $\{-1.5, -0.5, 0.5, 1.5\}$. As the equalizer algorithm has two inputs, we can prepare a graph showing c_n versus c_{n-1} from columns 5 and 6 of the table. Each of these points (pairs of c values) can be labelled, as shown by the letters in column 8. The resulting signal space is displayed in Fig. 3.17. In this figure points A, B, C, and D represent $d_n = -1$, whereas E, F, G, and H indicate $d_n = +1$. How can the space be partitioned into two regions ABCD and EFGH, representing $d_n = -1$ and $d_n = +1$?

Regions are needed because we are now going to assume that white noise exists in the channel. Thus, even if, say, point D is repeatedly sent, the signal space plot of received points will give a scatter of values about D. Generally, the probability density function of this noise will fall off rapidly with distance from point D. There will even be (hopefully rare) occasions when the received point lies in region EFGH, causing a received error.

The maximum likelihood partition boundary consists of the locus of points that are equidistant from corresponding members of the two point sets, as shown in Fig. 3.17. This boundary will minimize errors due to channel noise.

Any linear equalizer will appear as a straight line on this graph. An example of such a threshold is shown. It is easy to see that it is impossible to produce the maximum likelihood boundary with a linear equalizer (a single straight line). Thus, the best equalizer is nonlinear.

The above result can be improved by increasing the number of regions. Four regions are used (as shown in Fig. 3.18) along with their maximum likelihood boundaries. By referring to Table 3.1, it is seen that region AB corresponds to $d_n, d_{n-1} = -1, +1, \ldots$

An important feature of this figure concerns transitions when a new channel output occurs. Since the input is binary, movement is only possible to two of the four regions. The other two regions are 'off

Table 3.2 Next state and output table for $d_n = 1,\ -1$, given a noiseless channel

	$n = 1$	$n = 2$	$n = 3$
Present signal space point	D	H	B
Channel input data $d_n d_{n-1}$	$-1, -1$	$+1, -1$	$-1, -1$
$c_n c_{n-1}$	$-1.5, -1.5$	$0.5, -1.5$	$-0.5, 0.5$
New data d_n	1	-1	—
Next signal space point	H	B	

Table 3.3 Channel output of Table 3.2, with added noise

	$n = 1$	$n = 2$	$n = 3$
Noisy c_n, c_{n-1}	$-0.6, -1.5$	$0.1, -0.6$	$0.4, 0.1$
Point	R	S	T

limits'. This limited travel feature can be exploited when correcting a noisy signal. We will illustrate the process by a numerical example.

Assume that the channel output is presently at point D in Fig. 3.17, and that two new inputs $+1, -1$ are applied. Using Table 3.1, Table 3.2 can be constructed, assuming no channel noise. The shift of data from one column to the next is shown. The next signal space point is found from Table 3.1. For example, the transition from D to H (given the new input is $+1$) is shown in the last two columns of row 9 of Table 3.1.

Next, assume lots of channel noise is added to the $c_n c_{n-1}$ row of Table 3.2, giving the results in Table 3.3. Next, an ideal four-region signal space is constructed. Each region contains a single ideal point which is at the centroid of all actual points in the region. The reason for

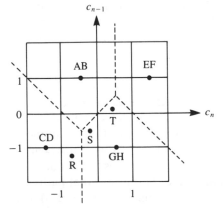

Figure 3.19 Ideal point four-region signal space, with noisy received points R, S and T

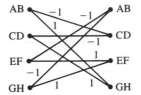

Figure 3.20   State transition trellis for the example

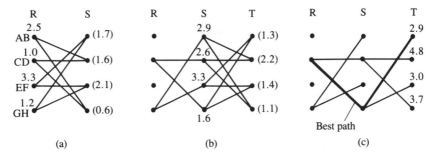

Figure 3.21   Two stages of the Viterbi algorithm

this is that shortly distance calculations will be made, and the centroid point simplifies this process. Figure 3.19 shows the four-region space, the centroid point of each region, and the three noisy received points R, S, T of Table 3.3. Thus the sequence of received regions is CD, GH, GH.

A trellis diagram is also needed, showing the allowed transitions from region to region. Working from Table 3.1, the first row shows that $c_{n-1}c_{n-2} = D$ has a transition to $c_n c_{n-1} = D$. Using this transition and those from the other rows, Fig. 3.20 is formed as shown. This shows that a transition from GH to GH is not permitted, verifying that the received sequence is noisy!

In order to recover the original data, the Viterbi algorithm can be used (see Roden (1992) and later in this book). To start, the Euclidian (or radial) distance from R to the ideal region points (CD, AB, etc.) is found. We repeat the process for S, and then form the two-stage trellis diagram of Fig. 3.21(a).

Obviously we would like this distance to be as small as possible. Consider the point CD in column S. If path AB, CD is followed, the summed distance is $2.5 + 1.6 = 4.1$, whereas path CD, CD has distance $1.0 + 1.6 = 2.6$. The total distance error metric will be the sum of the distances associated with the various paths of Fig. 3.21(a). The essence of the Viterbi algorithm is to eliminate all arriving paths to the nodes in column S except for the shortest distance one. This means that path AB, CD can be deleted. In general, all points in column S in Fig. 3.21(a) will have only one arrival path.

This leads to the simplification in Fig. 3.21(b), which also shows the

addition of point T to the trellis. The Viterbi algorithm is now applied
to the ST trellis, giving Fig. 3.21(c). Finally, the lowest distance path is
identified as CD, GH, AB, which turns out to be the correct sequence.
From this sequence, the original data can be identified by using Table
3.1.

The realization of such an equalizer is straightforward with a digital
signal processing (DSP) device. At each stage the radial distances from
the noisy received point to the ideal received points must be calculated.
Then the survivor transitions are found and stored, and the total
distances updated. After the complete trellis has been traversed, the
lowest distance path is found by working through the trellis from right
to left. Finally, this information can be used in conjunction with Table
3.1 to make the best guess for the noisy input sequence. It must be said
that long sequences will need a lot of memory storage and processing
time.

Several observations can be drawn from this simple example. Linear
equalizers are certainly not optimal. If the received signal space is
partitioned into more regions, knowledge of permitted transitions
between these regions can be used (with the Viterbi algorithm) to make
a 'best guess' of the transmitted data. Four regions were used in the
above example for simplicity, but even greater accuracy could be
achieved by using eight regions, one for each of the points in Fig. 3.18.

3.3 Conclusions

This chapter has looked at various schemes for system equalization. In
Sec. 3.1, analogue frequency domain concepts were employed. By
using amplitude and group delay correction networks, it was found
possible to create an excellent combined passband response. The order
of correction was found to be important. It is generally best to correct
for group delay distortion first, then the amplitude distortion. If done
the other way round, the amplitude correction network generates a large
peak of group delay near the passband edge, which sets a high overall
passband group delay level when the group delay correction is in place.
This leads to large waveform delay, which can cause problems in some
applications.

In Sec. 3.2 the discrete time domain was used for equalization, the
accent being on i.s.i. minimization. Both zero forcing and least squared
error criteria were explored, with the latter offering a slightly better
result. The section finished with a more complicated method for symbol
detection, based on nonlinear decision boundaries and the Viterbi
algorithm.

There are therefore many equalization methods available. The
tendency these days is towards robust digital signal processing

algorithms, since these algorithms are easily implemented in VLSI, or in software. It seems that there is still much research work to be done in this area, to achieve a given blend of speed, accuracy, convergence time, and cost.

References

Carlson, A. B. (1975) *Communication Systems*, McGraw-Hill Book Company, New York.

Cowan, C. F. N. (1992) 'Communications equalisation using adaptive techniques', Circuit Theory and DSP Colloquium, *IEE Digest*, no. 1992/037 (Nonlinear equalizer design).

Fryer, M. J. and Greenman, J. V. (1987) *Optimisation Theory*, Edward Arnold, London, pp. 33–37 (Lagrange multipliers).

Haykin, S. (1986) *Adaptive Filter Theory*, Prentice-Hall, Englewood Cliffs, New Jersey (LMS algorithm).

Lucky, R. W., Salz, J. and Weldon, E. J. (1968) *Principles of Data Communication*, McGraw-Hill Book Company, New York (Equalizers).

Oppenheim, A. V. and Schafer, R. W. (1975) *Digital Signal Processing*, Prentice-Hall, Englewood Cliffs, New Jersey.

Press, W. H., Flannery, B. P., Teukolsky, S. A., and Vetterling, W. T. (1986) *Numerical Recipes*, Cambridge University Press, Cambridge, U.K.

Rhodes, J. D. (1976) *Theory of Electric Filters*, John Wiley and Sons, Inc., New York (New analogue design).

Roden, M. S. (1982) *Digital and Data Communication Systems*, Prentice-Hall, Englewood Cliffs, New Jersey (Viterbi decoding).

4 Modulation for band-pass channels

LARRY LIND

4.1 Introduction

Bandpass channels exist in a variety of forms. There are telephone channels with a bandwidth of say 0.4 to 3.4 kHz, radio and TV channels with a variety of bandwidths, satellite communication channels with larger bandwidths, microwave channels with even larger bandwidths, and optical fibres with huge bandwidths. Sometimes the bandwidths are limited by the physical parameters of the medium. Examples of such limitations are loss and the effect of loading coils in copper pair cables, and the molecular resonances that absorb power in optical fibres and in the atmosphere.

There are other instances where the bandwidth is limited by regulatory constraints. This is true in all cases of radio frequency broadcasting, including not only entertainment radio and TV, but also citizens' band users, garage door openers, and mobile phones. There is a voracious and increasing need for more and more channels in the radio frequency spectrum. Often, crystal-controlled oscillators and sharp filters are used to ensure that the regulatory conditions for centre frequency and bandwidth are met.

In the past, many of these bandlimited channels were used for analogue communication, often speech. Today, however, there is an increasing tendency to send data down these channels. There are various issues concerned with data transmission over a bandlimited channel. How much bandwidth is needed? How can the transmission be achieved in a reliable manner? What waveform shapes should be used? How can the receiver differentiate between these shapes? What international standards exist for modulated data transmission? We address these issues in the following sections.

Basically, however, the design of band-pass channels for data transmission can be viewed as an extension of the principles of baseband data transmission, described in Chapter 1. In this chapter we will make extensive use of the signal space concept, which enables the transmitted waveforms to be represented by a constellation of points. Simple geometric arguments can then be used to explore the 'efficacy' of a constellation. A method for producing good constellations will be given towards the end of the chapter.

4.2 Background

The task at hand is to send an isochronous data stream over a band-pass channel. Let each symbol of the data stream last for T seconds, which will be called the element period. A common symbol is a binary data bit, having just two values. However, the symbol can be more complicated, and contain, say, m levels or values. Later it will be seen that a multilevel symbol can carry more information during the element period, but at the cost of more susceptibility of error due to noise.

The type of waveform to be sent over the band-pass channel during an element period is usually assumed to be a portion of cosine wave. Its oscillation frequency f_c lies within the channel passband. There are several reasons for using a truncated cosine waveform. It is relatively easy to generate with analogue or digital circuitry. Also, its spectral content is easy to find. Let an element period of cosine wave be represented by:

$$g(t) = \cos(2\pi f_c t) \, \text{rect}(t/T)$$

where T is the element period.

The Fourier transform of this waveform is

$$G(f) = 0.5[\delta(f - f_c) + \delta(f + f_c)]*[T.\text{sinc}(fT)]$$
$$= 0.5T[\text{sinc}((f - f_c)T) + \text{sinc}((f + f_c)T)]$$

where the asterisk in the first line indicates convolution.

To illustrate this result, let $f_c = 10\,\text{kHz}$. Also, let the element period (e.g., the square wave pulse envelope waveform in Fig. 4.2) consist of exactly 20 cycles of cosine waveform. Then $T = 20/f_c = 2\,\text{ms}$, giving a transmission rate of 500 baud. The amplitude spectrum for this timelimited cosine wave is shown in Fig. 4.1.

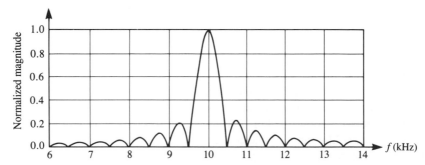

Figure 4.1   Amplitude spectrum of a 20 cycle cosine waveform

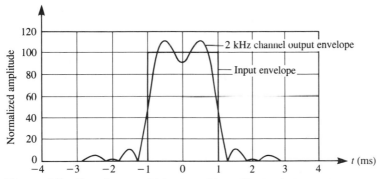

Figure 4.2   Envelopes of ideal and bandlimited channels

The band-pass channel should pass a reasonable amount of this spectrum. Thus, let the channel have a rectangular response, from 9 to 11 kHz. Then the three strongest lobes will be passed, but all others suppressed.

To gain an idea of how this bandlimiting channel distorts and spreads the input waveform, we consider the *envelope* of the modulating cosine data signal, which is a rectangular time domain waveform, with a width of 2 ms. This envelope is then applied to the low-pass equivalent of the channel, which is modelled as an ideal low-pass filter with upper cut-off frequency of 1 kHz. The channel output envelope of the 10 kHz carrier is shown in Fig. 4.2. Also shown is the input envelope.

The channel stopband gives rise to distortion within the element period, as well as spillover into the adjacent element periods. This spillover effect (or i.s.i.) must be borne in mind when recovering baseband data from the incoming modulated waveform.

In Fig. 4.2 the channel seems to be supplying an output before the input occurs. This is one of those effects that arises in ideal communication theory, but cannot occur in practice! The reason for the anticipatory output is because the channel is assumed to have zero group delay, in the production of Fig. 4.2. All real channels will have finite group delay. This delay gives rise to a time shift between input and output. The channel output should actually be to the right in Fig. 4.2 to reflect this fact; then the output will start to change only after the input is applied.

Figure 4.2 shows that some i.s.i. will exist on the output envelope, especially at the ±1 ms sample points. A better response could be obtained if the input plus channel (plus equalizer) has a 100 per cent raised cosine frequency spectrum. The output waveform envelope would then be given by Eq. (1.3d) (Chapter 1) and have the appearance of Fig. 1.7(d), which not only has zero i.s.i., but also good immunity to timing jitter.

The channel is assumed to be linear, thus the isochronous data stream, composed of 20 cycle cosine wave pieces, will have a spectrum given by the superposition of the element period spectra. These spectra add in a complicated fashion, for their phase as well as amplitude responses must be taken into account in adding them together. However, regardless of the peak amplitude or starting phase, all sinusoid 2 ms pieces will have an amplitude response of the form as shown in Fig. 4.1, which demonstrates that the 2 kHz channel is well suited to this type of waveform.

4.3 Symbols

A symbol is defined as a waveshape that is an element period long. In the following treatment a symbol set is formed whose members are all sinusoids at the common frequency f_c, lasting for T seconds. The differences between members of the set are their peak amplitudes and starting phases.

It is now convenient to describe a graphical method of representing a symbol (Sklar, 1988). A signal space plot is made to this end, as shown in Fig. 4.3. If the starting time for the symbol period is regarded as a time origin, then any sinusoid can be decomposed into the summation of a sine wave and cosine wave:

$$A \sin (2\pi ft + \phi) = A \cos \phi \sin (2\pi ft) + A \sin \phi \cos (2\pi ft)$$

This sinusoid can be plotted as a point in the two-dimensional signal space of Fig. 4.3(b). The horizontal and vertical components are $A \cos \phi$ and $A \sin \phi$. The point has a radial distance A from the origin, and an angle ϕ with respect to the positive horizontal axis (measured in an anticlockwise direction). This angle is the starting phase of the sinusoid.

Both axes of the signal space are real. Signal space does not show the element period, or the sinusoid over this period. Instead, it concentrates on the features of amplitude and starting phase. A collection of symbols therefore has a convenient representation in signal space. The resulting point set is sometimes called the *constellation* for a particular modulation

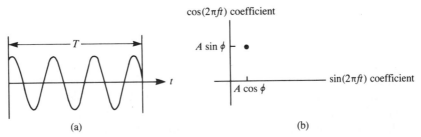

Figure 4.3 Relation between a symbol and its plot in signal space: (a) time waveform, (b) equivalent point in signal space

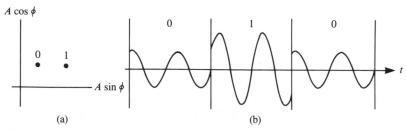

Figure 4.4 (a) A two-point constellation, and (b) resulting time waveform for the data sequence 010

scheme. By way of example, a two-point (binary) constellation is shown in Fig. 4.4(a). One possible transmitted time waveform using this constellation is shown in Fig. 4.4(b). In general, if n bits of data are required to be sent during a single element period, a constellation of 2^n points is needed.

The channel will not only attenuate the transmitted signal, but also produce some additive noise. Thus, the sine and cosine components of a received waveform will give a point that does not always lie exactly at an ideal signal point. The receiver circuitry does the best job possible in calculating the sine and cosine components in the received waveform, over the element period. One way to do this is to multiply the received signal by cosine and sine versions of the carrier waveform, and then to integrate the products over the element period. What emerges are effectively the two Fourier coefficients of the received waveform, at the carrier frequency. These coefficients are then plotted as a point in signal space. The decision as to which point was sent is very easy to make—we choose the closest ideal point, and declare it the sent point, this strategy being called *maximum likelihood estimation* theory.

4.4 Types of constellations

Examples of constellation types are given by Sklar (1988) and Peebles (1987).

4.4.1 Amplitude shift keying

An eight-point constellation for an *amplitude shift keying* (ASK) scheme is shown in Fig. 4.5a. Each point in this signal space represents a binary pattern. With eight points, it is possible to send three bits per element period. If the starting phase for each symbol sinusoid is zero, all signal space points then lie on the horizontal axis.

Since the radial distance from the origin is the amplitude of the sinusoid, the squared distance is proportional to power. It is a common

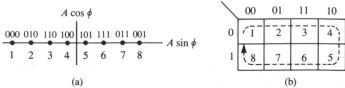

(a)   (b)

Figure 4.5   (a) An eight-point constellation for amplitude shift keying (ASK) and (b) a Karnaugh map walk, to determine the labelling of points in (a)

requirement to minimize the mean power of a constellation. If each signal space point is viewed as a point mass of unit weight, then the squared distance to each point mass is proportional to its moment of inertia. We know from mechanics that if the net moment of inertia of a collection of point masses is to be minimized (i.e., the power is to be minimized), then the origin should be placed at the *centroid* of the point mass system. This has been done with the constellation of Fig. 4.5.

It is necessary to have some space between points in order to guard against additive system noise, as will be seen later. We next assume that the noise scatter around each received signal space point is the same (regardless of the point), and that the associated noise probability density function has circular symmetry. Now the spacing between points is a trade-off between error performance and the transmitted power. The same trade-off should exist between all pairs of nearest neighbour points, which means that the points should be equally spaced. For an equally spaced ASK constellation, expressions can be found for the peak power P_p and the mean power P_m. An n-point layout (with the origin at the centroid) is assumed, with unit spacing between points. For n odd, the two points furthest from the origin have distance $(n - 1)/2$, giving the peak power as:

$$P_p = 0.25(n - 1)^2 \tag{4.1}$$

The mean power for n odd is calculated as:

$$P_m = \frac{2}{n} \sum_{k=1}^{(n-1)/2} k^2$$

For K terms, the summation reduces to:

$$\sum_{k=1}^{K} k^2 = \frac{K(K + 1)(2K + 1)}{6} \tag{4.2}$$

then P_m is given by the simple expression:

$$P_m = \frac{(n^2 - 1)}{12} \tag{4.3}$$

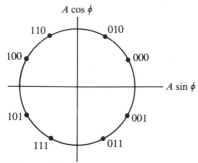

Figure 4.6   An eight-point constellation for phase shift keying (PSK)

When similar calculations are performed for n even, curiously the same results of Eqs. (4.1) and (4.3) are obtained. Thus they apply for all ASK constellations, given the above assumptions.

Sometimes the peak power is more important than the mean power. For example, the transmitter might have a power amplifier that is destroyed if the peak power is exceeded!

One good labelling scheme for ASK is shown in Fig 4.5(a), where a 'grey' code is used. The Karnaugh map has the property that only one bit changes between adjacent cells. To achieve this property, its axes are each given by a 'grey' code sequence. The ASK constellation point labels are produced from a walk on the Karnaugh map, as seen in Fig. 4.5(b). If noise does cause a wrong amplitude to be detected, it will most probably be detected as a point adjacent to the sent point, and the grey code labelling will only produce a single bit output error for such a case.

ASK is simple to generate and detect. However, it does not make good use of signal space. Only one dimension is used; the other is completely ignored! Quantitative results on this modulation scheme, and the ones in Secs 4.4.2 and 4.4.3, are given in the references at the end of the chapter.

4.4.2 Phase shift keying

The main feature of *phase shift keying* (PSK) is that all points in the constellation have the same amplitude. The only difference in the related sinusoids is that each one has a different starting phase. All signal space points lie on a circle of radius R (which will be found shortly). Since all points have the same amplitude, the transmitted waveform will have (ideally) a constant envelope. It makes good sense to use this form of modulation. The transmitter is constantly run at full power, implying that the receiver always has the maximum possible received signal-to-noise ratio. If PSK were demodulated to a sequence of baseband signals, each signal space point would be demodulatd to a complex number (representing a level) of fixed magnitude but variable phase.

An example of an eight-point PSK constellation is given in Fig. 4.6. The points have been labelled with the same Karnaugh walk as shown in Fig. 4.5(b). Whereas for the ASK labelling it was not necessary for the walk to be re-entrant (cell 8 adjacent to cell 1), it is necessary here. There are many possible re-entrant walks on a three-variable Karnaugh map, each producing a distinct labelling scheme. Therefore, there are many best labelling schemes that can be employed. Since eight points are used, there is again the transmission of three binary bits of data per element period.

The radius of the PSK points can be found by using Fig. 4.7. The angle ϕ is:

$$\phi = 2\pi/(2n)$$

Then $R = 1/\sin \phi$, and the peak and mean powers are given by:

$$P_{\mathrm{p}} = P_{\mathrm{m}} = 1/\sin^2 (\pi/n) \tag{4.4}$$

For large n, $\sin(\pi/n) \approx \pi/n$, and this expression is approximately:

$$P_{\mathrm{p}} = P_{\mathrm{m}} \approx (n/\pi)^2 \tag{4.5}$$

This digital modulation method is generally less efficient than ASK from the mean power point of view, since no points are allowed inside the circle. For example, an ASK 64-point constellation requires $P_{\mathrm{m}} = 341.2$, whereas the PSK equivalent needs $P_{\mathrm{m}} = 415.3$, and increase of 22 per cent mean power. However, the peak power criterion favours PSK; for 64-point ASK, $P_{\mathrm{p}} = 992.2$, compared to the PSK value of $P_{\mathrm{p}} = 415.3$.

Figure 4.6 shows that there is a large area inside the circle which is not being used. This, from a signal space point of view, is wasted space. The next modulation scheme utilizes signal space in a much more efficient fashion.

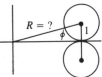

Figure 4.7   A construction used for finding the peak amplitude of all signal space points

4.4.3 Quadrature amplitude modulation

With *quadrature amplitude modulation* (QAM), both phase and amplitude vary from one signal space point to the next. In a sense, both ASK and PSK can be considered as special cases of QAM. Because of their specialized nature though, we continue to distinguish them by using distinctive acronyms.

A common QAM constellation is formed by using a rectangular or

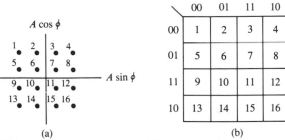

(a)   (b)

Figure 4.8   (a) A 16-point QAM constellation and (b) a Karnaugh map labelling scheme

square grid of points, a power of two in number. A 16-point layout is displayed in Fig. 4.8(a). Each point is separated from its nearest neighbours by a spacing of one unit. Signal space is now filled in a rather compact fashion. One would anticipate much smaller peak and mean powers. For an $m \times n$ array of points (both m and n are even numbers), the mean power is found from:

$$P_m = \sum_{l=1}^{m/2} \sum_{k=1}^{n/2} (k - 0.5)^2 + (l - 0.5)^2$$

Using the ASK results above, P_m reduces to the simple form:

$$P_m = \frac{(m^2 + n^2 - 2)}{12} \tag{4.6}$$

The peak power is given by:

$$P_p = \frac{(m - 1)^2 + (n - 1)^2}{4} \tag{4.7}$$

Continuing with the example of an 8×8 (64-point) constellation, the mean and peak powers for QAM are $P_p = 10.5$, $P_m = 24.5$, which represent large reductions compared to the ASK and PSK results.

The labelling scheme for QAM is again based on a Karnaugh map. An example is given in Fig. 4.8(b). The meaning is that in Fig. 4.8(a), point 1 would be labelled by the four-bit pattern 0000, point 2 by 0100, and so on.

Two internationally agreed layouts are shown in Fig. 4.9 (there are many others in the CCITT V-series). The first one is from the V.26 recommendation, which allows data to be sent at a 24 kbit/s rate with a carrier frequency of 1800 Hz. It is an example of four-phase PSK. The second layout is more ambitious, and comes from the V.29 recommendation. The data rate is now faster (at 9.6 kbit/s) using a carrier frequency of 1700 Hz. The points are relatively far apart with respect to phase shift, and so this layout is especially good for channels with phase jitter. However, for a given error performance (spacing

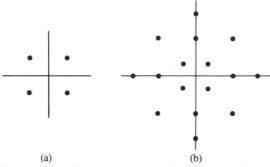

(a)   (b)

Figure 4.9 Some common signal space layouts: (a) the V.26 PSK constellation and (b) the V.29 QAM constellation

between points), the V.29 constellation needs more average and peak power than the V.26 one, and so needs a better signal-to-noise ratio.

The two layouts show that by increasing the number of points in a constellation, the *bit* data rate can be increased while keeping the symbol data rate relatively slow. There are examples from digital radio where a constellation of 1024 points is used, which results in 10 bits per element period being sent.

4.4.4 Optimal constellations

The problem of finding a constellation that minimizes the peak and mean powers, given that each point has unit spacing to its nearest neighbours, is equivalent to arranging coins on a plane surface to be as close together as possible.

We introduce the notion of perfect constellations, which can be formed as follows. The first one is just a single coin (of radius 0.5 units), with its centre at the origin. This constellation will be called C_0. The next perfect constellation, C_1, is formed by surrounding C_0 with a hexagon of one coin per side, as shown in Fig. 4.10(a).

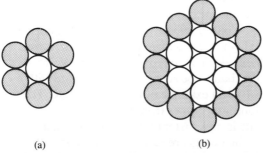

(a)   (b)

Figure 4.10 (a) The C_1 constellation. (b) The addition of another hexagonal ring produces the C_2 constellation

C$_2$ is formed by adding a new hexagon which fits around the points in C$_1$. This hexagon consists of two coins per side. In general, C$_n$ is formed from C$_{n-1}$, with the addition of a hexagon of n coins per side. It can be shown that the total power P_n of C$_n$ is given by the recursive relation:

$$P_n = n(n + 1)(5n^2 + 5n + 2)/4 \tag{4.8}$$

Each of these perfect constellations represents a close-packed structure, with the maximum possible symmetry.

The number of points N_n in C$_n$ is easily found to be:

$$N_n = 1 + 6 \cdot 1 + 6 \cdot 2 + \ldots + 6 \cdot n = 3n(n + 1) + 1 \tag{4.9}$$

The first few results from this equation are $N_1 = 7$ (Fig. 4.10a), $N_2 = 19$ (Fig. 4.10b), $N_3 = 37$, $N_4 = 61$, and $N_5 = 91$.

Unfortunately, these numbers are not very useful for practical designs. Usually the requirement is to have a 'power of two' number of points. To meet this requirement, *part* of the outermost hexagon is usually discarded. For example, a 16-point constellation can be formed by using C$_2$, with three coins removed. The question is, which three should be discarded?

To answer this question, assume that an n-coin constellation exists that has minimal power. Therefore the origin has been placed at the centroid of the constellation. In the x-direction, the constellation has the power:

$$P_{n,x} = \sum_{i=1}^{n} x_i^2 \tag{4.10}$$

and the mean:

$$\mu_{n,x} = \frac{1}{n} \sum_{i=1}^{n} x_i \tag{4.11}$$

We want to add another coin from the outer hexagon to the constellation, such that the increase in power is minimized. When adding the new point, the centroid will shift to:

$$\mu_{n+1,x} = \frac{1}{n+1} \sum_{i=1}^{n+1} x_i = \frac{x_{n+1}}{n+1} \tag{4.12}$$

The new power in the constellation about $\mu_{n+1,x}$ is given by:

$$P_{n+1,x} = \sum_{i=1}^{n+1} (x_i - \mu_{n+1,x})^2 \tag{4.13}$$

If this sum is split into a sum from 1 to n and the last term is treated separately, some algebra leads to the result:

$$P_{n+1,x} = P_{n,x} + [n/(n + 1)]x_{n+1}^2 \tag{4.14}$$

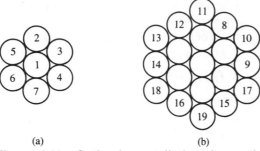

(a)   (b)

Figure 4.11 Optimal constellations for partially filled hexagons. (a) C_0 to C_1 evolution and (b) C_1 to C_2 evolution

In the y-direction the same reasoning can be applied, with the overall result:

$$P_{n+1} = P_n + [n/(n+1)](x_{n+1}^2 + y_{n+1}^2) \qquad (4.15)$$

Obviously, the new point should be chosen to minimize this equation. Hence, the best outer hexagon point to add is simply the one closest to the present centroid. The new centroid is then computed, and the process repeated. At each stage the result must be optimal, since the additional power is constantly being minimized. Sometimes, because of symmetry, there will be a tie for the new point to be added. Then one of the best points is arbitrarily selected. Figure 4.11 shows the result of this optimization process, point by point, for the C_1 and C_2 constellations.

This strategy can also be worked in reverse. For example, if a 16-point layout is desired, one can start with the C_2 constellation, consisting of 19 points. Points are then removed successfully, according the rule: 'the best outer hexagon point to subtract is simply the one furthest from the present centroid'. The new centroid is then computed, and the process repeated. At each stage the result must be optimal, since the reduced power is constantly being minimized.

It does not matter whether we follow an additive or subtractive approach with the outer ring; both methods give the same final result. To minimize the work, however, it is better to use the additive method if the outer ring is mainly empty, and the subtractive approach if the outer ring is mainly full of signal space points.

The best 16-point layout is found by the subtractive approach, and is shown in Fig. 4.12. Using the centroid as the origin, the mean power in this constellation is $P_{mo} = 2.1875$. An equivalent 16-point QAM grid has the mean power $P_{mq} = 2.50$. Thus the optimal arrangement gives a mean power reduction of:

$$10 \log_{10} (P_{mq}/P_{mo}) = 0.58 \text{ dB} \qquad (4.16)$$

Labelling an optimal 16-point constellation is more difficult than the QAM scheme, because now a central point has six other points as

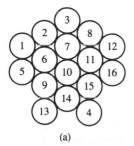

	00	01	11	10
00	1	2	3	4
01	5	6	7	8
11	9	10	11	12
10	13	14	15	16

(a)   (b)

Figure 4.12 Optimal 16-point constellation: (a) layout in signal space and (b) a Karnaugh map labelling for these points

nearest neighbours. One solution is shown in Fig. 4.12, along with an associated Karnaugh map. We note that this labelling has usually, but not always, the 'one-bit change between adjacent cells' property.

Should an optimal mean power configuration be used? The added complexity in the generation and reception equipment can be easily provided by the low-cost digital storage and computing devices that are now available. The power savings should make this type of constellation increasingly attractive to system designers.

4.5 Conclusions

This chapter has looked at some good ways of sending data over a modulated, bandpass channel. Various signal point constellation schemes have been explored, including one that appears to be optimal from a two-dimensional signal space packing point of view. This latter scheme requires more signal processing at the transmitter and receiver than a lattice QAM layout, but there is the promise of a modest reduction in peak signal power, average signal power, and hence a reduction in signal-to-noise ratio to achieve a given error probability. These layouts require complicated algorithms at the transmitter and receiver. However, with the digital signal processing power that is now available, there will be an increasing interest in inventing new signal space constellations that optimize some performance figure of merit (Forney and Wei, 1989), regardless of algorithm complexity.

References

Forney, G. D. and Wei, L. F. (1989) 'Multidimensional constellations—Part 1', *IEEE J. of Selected Areas in Comms.*, vol. 7, 877–92.

Peebles, P. Z. (1987) *Digital Communication Systems*, Prentice-Hall, Englewood Cliffs, New Jersey.

Sklar, B. (1988) *Digital Communications*, Prentice-Hall, Englewood Cliffs, New Jersey.

5 Making digital decisions

NIGEL J. NEWTON

5.1 Introduction

In Chapters 1 to 4 the problem of sending information in discrete (or digital) form over transmission channels of various kinds was introduced, along with ways in which this is often achieved in practice. The techniques used have evolved from the simple threshold detector at the end of an electrical cable detecting a low bit-rate binary signal to more complex systems using pulse amplitude modulation (PAM), sophisticated equalizers, self-retiming circuits, multiplexers, etc. The motive for these developments has been the need to make more efficient use of the transmission resources available. We might ask whether there is an *optimal* way to make use of these resources. Before attempting to answer this question we must define precisely what we mean by optimal; we must also bear in mind that the question may not be worth answering, as in practice we must always seek a compromise between technical optimality and cost. In addition, it may be that future developments in optical fibre networks reduce the need to squeeze every last drop of information-carrying capacity out of the physical equipment. Nevertheless, it is important for receiving equipment to make intelligent use of the raw information arriving over the transmission medium in deciding what is happening at the sending end. For example, it is common sense to use techniques such as pulse equalization to give good eye opening at the receiver's sampling instant, but are there ways of optimizing the design of such equalizers? The purpose of this chapter is to formalize some of these ideas and to introduce some general techniques in optimal decision-making. The techniques are illustrated on a few examples from digital transmission. The aim is not to give an exhaustive list of applications or potential applications, but to show that there are decision-making problems in transmission systems with optimal solutions which are either practical in themselves or which suggest practical, suboptimal solutions.

In digital transmission systems the receiver needs to estimate various features of the state of the transmitter, such as the symbols transmitted, the position of the frame alignment word (in time-division multiplex systems), the symbol timing reference, etc. These estimates have to be made on the basis of information available at the receiver, which, in the most general case, is the entire history of the incoming continuous-time waveform. However, it may be a requirement that the estimates be made on the basis of less information than this, for example from continuous valued samples of the incoming waveform taken at integral multiples of the symbol period, or even quantized versions of these samples. Since the information available at the receiver is corrupted by noise and possibly by the effects of other random phenomena, the design of good receiver strategies is a problem in statistical estimation.

Typically, we want to estimate a parameter, a, on the basis of an observation, Y, which is perturbed by some random phenomenon. (In what follows random variables will be denoted by upper-case letters and nonrandom quantities by lower-case letters.) The parameter and observation may be *discrete*, *continuous* or even *function-valued* in various combinations. For example, we may wish to estimate the symbols transmitted (discrete) on the basis of the whole history of the noise-corrupted signal arriving at the receiver (function valued). If the parameter to be estimated is discrete then the problem is usually referred to as a *decision problem* rather than one of estimation. We will look in detail at decision problems only. Decision and estimation techniques for communication engineering are developed in depth by Cattermole (1986).

It may be possible to regard the parameter as being random in its own right and as having a *prior distribution* (which means that we have some notion of the likely values of the parameter before we 'see' the observation), in which case this can be incorporated into the decision-making process. For example, if it is known that some symbols are a priori more likely to be sent over a digital link than others then this can be taken into account by the receiver strategy.

If the parameter is discrete, $a \in \{a_1, a_2, \ldots\}$, and random with known prior distribution then Bayes' formula and its generalizations (Papoulis, 1972; Cattermole, 1986) can be used to find its observation-conditional *posterior distribution*:

$$P(A = a_i \mid Y) = \frac{p_{Y/A}(Y, a_i)}{p_Y(Y)} P(A = a_i) \tag{5.1}$$

Here $p_{Y|A}(y, a_i)$ is the $\{A = a_i\}$-conditional probability that $Y = y$ if Y is discrete, or the $\{A = a_i\}$-conditional density of Y evaluated at y if Y has a continuous distribution; $p_Y(y)$ (which is assumed to be nonzero) is, correspondingly, the probability that $Y = y$ or the density of Y evaluated at y. There is a clear-cut best decision criterion for A called the *maximum a posteriori probability* (MAP) criterion:

$$A^* = \arg \max_a P(A = a \mid Y) \tag{5.2}$$

This picks the value of a which maximizes $P(A = a|Y)$, the value which has the highest a posteriori probability of having occurred. The decision therefore has the minimum a posteriori probability of being wrong, i.e., it has the lowest probability of being wrong among all decision criteria which can 'see' only Y. If the set of possible values of the parameter is finite and the outcomes are a priori equally probable ($P(A = a_i) = 1/N$, for $i = 1, \ldots, N$) then the MAP criterion takes the form:

$$A^* = \arg \max_a p_{Y|A}(Y, a) \tag{5.3}$$

This criterion, called the *maximum likelihood* (ML) criterion is often used in contexts where A is nonrandom or has an unknown prior distribution.

5.2 Optimal detection of a binary random variable in the presence of a Gaussian disturbance

Suppose that A is a bipolar binary random variable with the prior distribution

$$P(A = +1) = p \tag{5.4a}$$

and

$$P(A = -1) = 1 - p \tag{5.4b}$$

and that the observation is given by:

$$Y = A + R \tag{5.5}$$

where R is a Gaussian random variable with mean zero and variance σ^2. Then, since $p_{Y|A}(y, a)$ is the Gaussian density with mean a and variance σ^2:

$$p_{Y|A}(y, a) = \frac{1}{(2\pi)^{\frac{1}{2}}\sigma} \exp\left(\frac{-(y - a)^2}{2\sigma^2}\right)$$

it follows from Eq. (5.1) that:

$$P(A = +1 \mid Y)$$
$$= \frac{p \exp\left(-(Y - 1)^2/2\sigma^2\right)}{p \exp\left(-(Y - 1)^2/2\sigma^2\right) + (1 - p) \exp\left(-(Y + 1)^2/2\sigma^2\right)} \tag{5.6a}$$

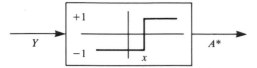

Figure 5.1 The maximum a posteriori (MAP) decision

and:

$$P(A = -1 \mid Y)$$
$$= \frac{(1 - p) \exp\left(-(Y + 1)^2/2\sigma^2\right)}{p \exp\left(-(Y - 1)^2/2\sigma^2\right) + (1 - p) \exp\left(-(Y + 1)^2/2\sigma^2\right)} \quad (5.6b)$$

The MAP criterion decides in favour of the value of A which corresponds to the maximum of these two probabilities. In fact, since the natural logarithm, $\ln x$, is positive if $x > 1$ but negative if $x < 1$ then:

$$A^* = \operatorname{sgn} \ln \left[\frac{P(A = +1 \mid Y)}{P(A = -1 \mid Y)} \right]$$

$$= \operatorname{sgn} \ln \left[\Lambda_{1,-1} \frac{p}{1 - p} \right]$$

$$= \operatorname{sgn} \left(\ln(p) - \ln(1 - p) + 2Y/\sigma^2 \right) \quad (5.7)$$

where sgn x is the sign function and

$$\Lambda_{1,-1} = \frac{p_{Y|A}(Y, 1)}{p_{Y|A}(Y, -1)} \quad (5.8)$$

is the so-called *likelihood ratio*. Thus the MAP criterion can be implemented by means of a simple threshold detector acting on Y with threshold:

$$x = \frac{\sigma^2}{2} \left(\ln(1 - p) - \ln p \right) \quad (5.9)$$

This is illustrated in Fig. 5.1.

The threshold, x, is the point at which

$$p n(1, \sigma^2) = (1 - p) n(-1, \sigma^2) \quad (5.10)$$

where $n(\mu, \sigma^2)$ is the Gaussian density with mean μ and variance σ^2. This point is illustrated in Fig. 5.2.

The probability of a decision error by the MAP criterion is given by:

$$P_e = E \min_a P(A = a \mid Y)$$

$$= \int \min_a p_{Y|A}(y, a) P(A = a) \, dy \quad (5.11)$$

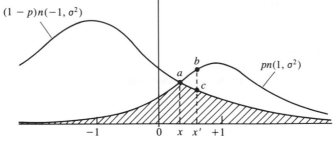

$(1 - p)n(-1, \sigma^2)$

$pn(1, \sigma^2)$

b
a
c

-1 0 x x' $+1$

Figure 5.2   The optimal threshold

where E indicates mathematical expectation. This is the shaded area in Fig. 5.2. The consequence of an error in the threshold setting can be clearly seen in this figure: if it were moved to position x' the error probability would increase by the area of the triangle (abc). In the special case that $p = 1/2$, which corresponds to the maximum information content of the binary random variable, the optimal threshold is zero and the probability of error is

$$P_e = \int_1^\infty n(0, \sigma^2)(y) \, dy \qquad (5.12)$$

This is the error probability derived in Chapter 1 and illustrated in Fig. 1.4. (Here $a = 1$.) The distinction between the MAP and ML criteria can be seen clearly in the problem treated in this section—the ML criterion decides on sgn Y as the value of A regardless of the value of the prior probability p, and is equivalent to the MAP criterion only if $p = 1/2$.

5.3   Optimal demodulation of a binary PAM signal in white Gaussian noise

Let A be the binary random variable of Sec. 5.2, and suppose that a pulse waveform $h(t)$ is amplitude-modulated by A. After transmission through a channel with impulse response $c(t)$, the signal is corrupted by the addition of white Gaussian noise, $R(t)$, with power spectral density σ^2 to yield the observation, $Y(t)$. The situation is that depicted in Fig. 5.3.

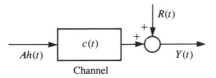

$R(t)$

$Ah(t)$ $c(t)$ $Y(t)$

Channel

Figure 5.3   A simple model for a transmission channel

The problem is to implement the MAP decision criterion for A given the observation $(Y(t), 0 \leqslant t < \infty)$. Suppose that the convolution of $h(t)$ with the channel impulse response is nonzero only in the interval $[0, \tau]$ for some $\tau < \infty$, and has unit 'energy', i.e.

$$g(t) = 0 \qquad \text{if } t < 0 \quad \text{or} \quad t > \tau \qquad (5.13a)$$

and

$$\int_0^\tau g^2(t) \, dt = 1 \qquad (5.13b)$$

where

$$g - h*\iota \qquad (5.14)$$

Generalizing Eq. (5.1) we might expect that:

$$\frac{P(A = +1 \mid Y(t), 0 \leqslant t \leqslant \tau)}{P(A = -1 \mid Y(t), 0 \leqslant t \leqslant \tau)} = \frac{p}{1 - p} \Lambda_{1,-1} \qquad (5.15)$$

where $\Lambda_{1,-1}$ is the likelihood ratio:

$$\Lambda_{1,-1} = \frac{p_{(Y(t), 0 \leqslant t \leqslant \tau)|A}(Y(t), 0 \leqslant t \leqslant \tau; 1)}{p_{(Y(t), 0 \leqslant t \leqslant \tau)|A}(Y(t), 0 \leqslant t \leqslant \tau; -1)} \qquad (5.16)$$

Unfortunately, the 'densities' in this expression are meaningless because $(Y(t), 0 \leqslant t \leqslant \tau)$ is a function-valued random quantity. However, this problem can be avoided if $Y(t)$ is represented by an infinite series of *orthonormal functions* in a manner similar to Fourier series:

$$Y(t) = \sum_{n=0}^{\infty} X_n f_n(t) \qquad \text{for all } 0 \leqslant t \leqslant \tau \qquad (5.17)$$

where the $f_n(t)$ are nonrandom functions such that:

$$\int_0^\tau f_n(t) f_m(t) \, dt = \begin{cases} 1 & \text{if } m = n \\ 0 & \text{otherwise} \quad \text{(orthonormality)} \end{cases} \qquad (5.18a)$$

and the X_n are random coefficients defined by the *projection* of $Y(t)$ onto the $f_n(t)$:

$$X_n = \int_0^\tau Y(t) f_n(t) \, dt \qquad (5.18b)$$

It is possible to choose the functions $f_n(t)$ so that the coefficients X_n are uncorrelated. (See Proakis (1983, Appendix 4A) or Cattermole (1986, Section 4.6).) This results in the so-called Karhunen–Loève series expansion for $Y(t)$. Equation (5.17) shows that the information carried by $(Y(t), 0 \leqslant t \leqslant \tau)$ is also carried by the sequence of random variables

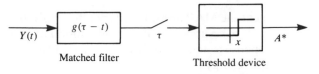

$$x = \sigma^2 (\ln(1 - p) - \ln p)/2$$

Figure 5.4 An implementation of the MAP criterion

(X_n); this is true since $(Y(t), 0 \leqslant t \leqslant \tau)$ can be found from (X_n). The multidimensional conditional densities of any finite sub-sequence of (X_n):

$$p_{(X_n, 0 \leqslant n \leqslant N)|A}(X_n, 0 \leqslant n \leqslant N; a)$$

are well defined and it turns out that their ratio has a limit as $N \rightarrow \infty$; this can be used to define the likelihood ratio $\Lambda_{1,-1}$. In fact:

$$\Lambda_{1,-1} = \lim_{N \rightarrow \infty} \frac{p_{(X_n, 0 \leqslant n \leqslant N)|A}(X_n, 0 \leqslant n \leqslant N; 1)}{p_{(X_n, 0 \leqslant n \leqslant N)|A}(X_n, 0 \leqslant n \leqslant N; -1)}$$

$$= \exp\left[\frac{2}{\sigma^2} \int_0^\tau g(t) Y(t) \, dt\right] \tag{5.19}$$

For details of the proof see one of the references cited previously.

The MAP decision criterion can now be found from Eqs. (5.15) and (5.19):

$$A^* = \text{sgn}\left(\ln p - \ln(1 - p) + \frac{2}{\sigma^2} \int_0^\tau g(t) Y(t) \, dt\right) \tag{5.20}$$

The integration in this expression is most easily implemented by means of a *matched filter*, which has impulse response $g(\tau - t)$. A suitable structure for an optimal receiver, which uses the MAP criterion, is shown in Fig. 5.4. In this, the output of the matched filter is sampled at time τ and the sample value is compared with the threshold x. Clearly a receiver using the ML criterion could be implemented in the same way but with a threshold of zero in the threshold device.

The signal entering the threshold device at time τ is given by:

$$Y = \int_0^\tau g(t) Y(t) \, dt$$

$$= \int_0^\tau A g^2(t) \, dt + \int_0^\tau g(t) R(t) \, dt$$

$$= A + R \tag{5.21}$$

where:

$$R = \int_0^\tau g(t) R(t) \, dt \tag{5.22}$$

Now R is Gaussian (see Theorem 4.6.5 in Cattermole (1986)) with mean zero and variance given by:

$$
\begin{aligned}
ER^2 &= \int_0^\tau \int_0^\tau g(s)g(t)ER(s)R(t)\, \mathrm{d}s\, \mathrm{d}t \\
&= \int_0^\tau \int_0^\tau g^2(t)\sigma^2\delta(t-s)\, \mathrm{d}s\, \mathrm{d}t \\
&= \sigma^2
\end{aligned}
\tag{5.23}
$$

Thus the problem reduces to that considered in Sec. 5.2 with the continuous valued observation, Y, obtained from the function-valued observation, $(Y(t), 0 \leqslant t \leqslant \tau)$, by means of a matched filter. In statistical terminology we say that Y is a *sufficient statistic* for making the MAP decision—the useful information in the whole stochastic process $(Y(t), 0 \leqslant t \leqslant \tau)$ can be carried in a single random variable, Y. This is perhaps surprising but very useful in that it corresponds well to the decision processes implemented in practice—the matched filter corresponds to a channel equalizer optimized for data recognition and the decision is made on the basis of the output of this filter at a single sampling instant (the instant of maximum eye opening discussed in Chapter 1).

5.4 Optimal demodulation of a binary PAM signal subject to intersymbol interference and additive white Gaussian noise

Section 5.3 described the optimal demodulator for a single pulse-amplitude-modulated bit in the presence of additive white Gaussian noise. The received waveform is passed through a matched filter, the output of which is sampled at time τ (the time beyond which the basic pulse shape at the receiver, $g(t)$, is zero). This is the instant at which, in the absence of any noise, the output of the matched filter reaches its peak. (See Fig. 5.5.)

Clearly, this technique can be used to demodulate a pulse-amplitude-modulated stream of bits with bit rate $1/\tau$ bit/s or less. The waveforms at the receiver bearing information about adjacent bits in the stream are then completely separated in time; this corresponds to a channel exhibiting no intersymbol interference (i.s.i). As discussed in earlier chapters, it is not economical to transmit bits at such a low rate—it is normal practice to use a transmission rate in excess of $1/\tau$ bit/s, and this leads to i.s.i. However, the optimal decision process in the presence of i.s.i. is very similar to that in its absence.

Suppose that (A_k) is a sequence of bipolar binary random variables

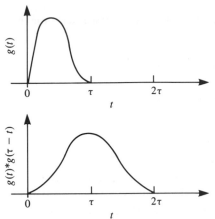

Figure 5.5 Behaviour of the matched filter

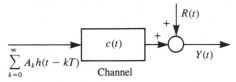

Figure 5.6 A simple model for a transmission channel

(the bit stream transmitted), and that a pulse waveform, $h(t)$, is amplitude-modulated by (A_k) at a rate of $1/T$ Hz. The resulting waveform is passed through a channel with impulse response $c(t)$, after which Gaussian white noise, $R(t)$, with power spectral density σ^2 is added. The situation is illustrated in Fig. 5.6.

Suppose that the convolution of $h(t)$ with $c(t)$, $g(t)$, has unit energy and is nonzero only in the interval $[0, (L + 1)T]$; i.e., the i.s.i. extends over L bit periods. There are various optimal criteria for deciding what information was sent, two of which are the *MAP sequence criterion* and the *MAP bit criterion*. The former makes a single decision on the entire sequence of bits sent, minimizing the probability that this decision is erroneous; the latter, on the other hand, makes separate decisions for each bit, minimizing the probability of bit error in each case. The difference between the two criteria is best illustrated by a simple example (involving no i.s.i.). Suppose that the output of a channel sampled at two successive bit periods is given by:

$$Y_0 = A_0 + R_0$$
$$Y_1 = A_1 + R_1 \tag{5.24}$$

where A_0 and A_1 are the binary values transmitted and the R_k are independent Gaussian variates with means zero and variances σ^2 (samples of some noise process). The problem is to estimate A_0 and A_1 given Y_0 and Y_1. Suppose that the prior probabilities are:

$$P(A_0 = -1, \quad A_1 = -1) = 0.1$$
$$P(A_0 = -1, \quad A_1 = +1) = 0.4$$
$$P(A_0 = +1, \quad A_1 = -1) = 0.4 \qquad (5.25)$$
$$P(A_0 = +1, \quad A_1 = +1) = 0.1$$

These probabilities might result, for example, from an encoding strategy which aims to ensure that there is an adequate number of level switchings for clock extraction. (See Chapter 7.) The MAP decision criterion for the *pair* (A_0, A_1) is:

$$(A_0, A_1)^* = \arg \max_{(i,j)} q_{i,j} \qquad (5.26)$$

where

$$q_{i,j} = P(A_0 = i, A_1 = j) \exp \left[\frac{-(Y_0 - i)^2 - (Y_1 - j)^2}{2\sigma^2} \right] \qquad (5.27)$$

(Here, the constant factor of the normal densities, $1/2\pi\sigma^2$, has been dropped.) That is $(A_0, A_1)^*$ maximizes the *joint* Y-conditional distribution of A_0 and A_1. However, the MAP criterion for A_0 *alone* is

$$A_0^* = \arg \max_i (q_{i,-1} + q_{i,1}) \qquad (5.28)$$

If $\sigma^2 = 0.1$, $Y_0 = 0.05$ and $Y_1 = 0.1$ then

$$q_{-1,-1} = 9.52 \times 10^{-7}$$
$$q_{-1,+1} = 2.81 \times 10^{-5}$$
$$q_{+1,-1} = 1.03 \times 10^{-5}$$
$$q_{+1,+1} = 1.91 \times 10^{-5}$$

and so

$$(A_0, A_1)^* = (-1, 1) \qquad (5.29)$$

but

$$A_0^* = 1 \qquad (5.30)$$

It is not easy to work directly with infinite sequences of bits as most, if not all, such sequences have zero prior probability of occurring; instead we consider finite sequences of N bits. As a further generalization of Eq. (5.1) we might expect that:

$$\frac{P(A_0 = a_0, \ldots, A_N = a_N \mid Y(t), 0 \leqslant t \leqslant NT)}{P(A_0 = b_0, \ldots, A_N = b_N \mid Y(t), 0 \leqslant t \leqslant NT)}$$
$$= \frac{P(A_0 = a_0, \ldots, A_N = a_N)}{P(A_0 = b_0, \ldots, A_N = b_N)} \Lambda_{a,b} \qquad (5.31)$$

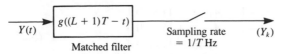

Figure 5.7   Derivation of sufficient statistics

where $\Lambda_{a,b}$ is the likelihood ratio of the two sequences $(a_0, \ldots, a_N)$ and $(b_0, \ldots, b_N)$. This can be proved to be so by methods similar to those referred to in Sec. 5.3 (see Proakis (1983, Sec. 6.3)), with:

$$
\Lambda_{a,b} = \exp\left[\frac{1}{\sigma^2}\sum_{k=0}^{N}(a_k - b_k)\int_0^{NT} g(t - kT)Y(t)\,dt \right.
$$
$$
\left. -\frac{1}{2\sigma^2}\sum_{j,k=0}^{N}(a_k - b_k)(a_j + b_j)\int_0^{NT} g(t - kT)g(t - jT)\,dt\right] \quad (5.32)
$$

The only terms in the likelihood ratio which involve the observation $Y(t)$ are the integrals:

$$
Y_k = \int_0^{NT} g(t - kT)Y(t)\,dt
$$
$$
= \int_0^{(k+L+1)T} g(t - kT)Y(t)\,dt \qquad \text{if } k \leqslant N - (L+1) \quad (5.33)
$$

and so the sequence (Y_k) constitutes sufficient statistics for the evaluation of $\Lambda_{a,b}$, i.e., the sequence (Y_k) contains all the information needed to make the MAP sequence decision. Since the posterior probabilities of individual bits in the sequence can be found from the posterior probabilities of the sequence (by summation of the probabilities over the other bits, as was done in the example above), the sequence (Y_k) contains all the information needed for the MAP bit decisions as well. The sequence (Y_k) can be evaluated by means of a matched filter whose output is sampled at the bit rate, as shown in Fig. 5.7. (Y_k is the output of the filter sampled at time $(k + L + 1)T$.)

We could continue by substituting the (Y_k) back into the expression for $\Lambda_{a,b}$, however we follow the development in Proakis (1983) and switch to an equivalent discrete time model.

Now

$$
Y_k = \int_0^{NT} g(t - kT)\sum_{j=0}^{N}A_j g(t - jT)\,dt + \int_0^{NT} g(t - kT)R(t)\,dt
$$
$$
= \sum_{i=-L}^{L} g_i A_{k-i} + R_k \qquad \text{if } k \leqslant N - (L+1) \quad (5.34)
$$

where (g_i) is the sampled autocorrelation function of $g(t)$:

$$
g_i = \int_{-\infty}^{\infty} g(t)g(t - iT)\,dt \quad (5.35)
$$

and (R_k) is a zero-mean Gaussian sequence with covariance

$$c_i = ER_k R_{k-i}$$
$$= \sigma^2 g_i \qquad (5.36)$$

The sequence (Y_k) can be transformed as follows in order to 'whiten' the noise (R_k). Let $G(z)$ be the two-sided z-transform of (g_i):

$$G(z) = \sum_{i=-L}^{L} g_i z^{-i} \qquad (5.37)$$

and let $A(z)$, $Y(z)$ and $R(z)$ be the z-transforms of (A_k), (Y_k) and (R_k), respectively. Since $g_{-i} = g_i$ the $2L$ zeros of $G(z)$ occur in reciprocal pairs, i.e., if x is a zero of $G(z)$ then so is $1/x$. $G(z)$ also has L poles at the origin of the z-plane, and so it can be factorized as follows:

$$G(z) = F(z)F(z^{-1}) \qquad (5.38)$$

where the zeros of $F(z)$ all lie inside (or on) the unit circle in the z-plane, and $F(z)$ has no poles (the poles of $G(z)$ all lie in $F(z^{-1})$), i.e., $F(z)$ is a polynomial of degree L whose roots lie inside (or on) the unit circle. For example:

$$-z^{-1} + 5/2 - z = \frac{1}{(2)^{\frac{1}{2}}} (1 - 2z) \times \frac{1}{(2)^{\frac{1}{2}}} (1 - 2z^{-1}) \qquad (5.39)$$

Now:

$$Y(z) = G(z)A(z) + R(z)$$
$$= F(z)\underline{Y}(z) \qquad (5.40)$$

where:

$$\underline{Y}(z) = F(z^{-1})A(z) + \underline{R}(z) \qquad (5.41)$$

and:

$$\underline{R}(z) = \frac{1}{F(z)} R(z) \qquad (5.42)$$

Since the sequence (Y_k) can be found from the sequence $(\underline{Y}_k)$ (the inverse transform of $\underline{Y}(z)$), the latter constitutes sufficient statistics for the MAP decisions. According to Eq. (5.41) the sequence $(\underline{Y}_k)$ is given by:

$$\underline{Y}_k = \sum_{i=0}^{L} f_i A_{k-i} + \underline{R}_k \qquad (5.43)$$

where f_i is the coefficient of z^i in $F(z)$. Also, since $(\underline{R}_k)$ is obtained from (R_k) by linear filtering, because:

$$E\underline{R}(z) = \frac{1}{F(z)} ER(z)$$
$$= 0 \qquad (5.44a)$$

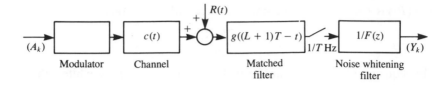

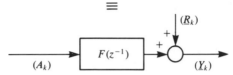

Figure 5.8   An equivalent discrete-time model

and since

$$ER(z)\underline{R}(z^{-1}) = \frac{1}{F(z)F(z^{-1})} ER(z)R^{-1}z^{-1})$$

$$= \sigma^2$$

(5.44b)

(R_k) is a white Gaussian sequence with mean zero and variance σ^2. (See Theorem 4.4.5 in Cattermole (1986).) For this reason the filter with transfer function $1/F(z)$, through which (Y_k) is passed in order to yield $(\underline{Y}_k)$, is called a *noise-whitening filter*. The combination of the modulator, the channel, the matched filter, and the noise whitening filter can now be represented by a simple discrete-time model, as shown in Fig. 5.8.

This model lends itself well to the discrete-time equalization techniques discussed in Chapter 3. It is important to reiterate that no useful information has been lost in replacing the process $(Y(t))$ by the samples $(\underline{Y}_k)$.

5.5 The Viterbi algorithm

Section 5.4 showed that the noise-whitened samples from the matched filter, $(\underline{Y}_k)$, form sufficient statistics for the MAP decisions. We now need to express the MAP sequence and bit decision criteria in terms of $(\underline{Y}_k)$, i.e., we seek the statistically optimal discrete-time equalizer.

If the binary symbols transmitted are statistically independent and a priori equally likely to be -1 or $+1$, then the MAP sequence criterion is identical to the ML sequence criterion, which decides in favour of the sequence maximizing the conditional density of $(\underline{Y}_k)$:

$$(\underline{Y}_k)/(\underline{A}_k)((\underline{Y}_k), (a_k)) = \prod_{k=0}^{N} n\left(\sum_{i=0}^{L} f_i a_{k-i}, \sigma^2 \right)(Y_k)$$

(5.45)

where $n(\mu, \sigma^2)$ is the Gaussian density with mean μ and variance σ^2. The sequence which maximizes the density above is also the sequence which minimizes the following *cost function* (because $\exp(-x)$, occurring in the Gaussian density, is a decreasing function, i.e., if $x > y$ then $\exp(-x) < \exp(-y)$):

$$J_N(a_0, \ldots, a_N) = \sum_{k=0}^{N} \left[\left(\sum_{i=0}^{L} f_i a_{k-i} \right)^2 - 2\underline{Y}_k \sum_{i=0}^{L} f_i a_{k-i} \right] \quad (5.46)$$

This sequence can be found by a recursive technique originally proposed as a means of decoding convolutional codes by A. J. Viterbi (1967), and adapted for sequence detection over channels with i.s.i. by G. D. Forney (1972); it is a special application of a more general procedure known as *dynamic programming*. Its application in the decoding of convolutional codes is described in Chapter 9.

For each of the 2^L possible sequences in the time interval $k - L + 1$ to k $(a_{k-L+1}, a_{k-L+2}, \ldots, a_k)$ let

$$V_k(a_{k-L+1}, a_{k-L+2}, \ldots, a_k) = \min_{a_0, a_1, \ldots, a_{k-L}} J_k(a_0, \ldots, a_k) \quad (5.47)$$

and:

$$M_k(a_{k-L+1}, a_{k-L+2}, \ldots, a_k) = (a_0^o, a_1^o, \ldots, a_{k-L}^o) \quad (5.48)$$

the right-hand side of Eq. (5.48) being the sequence which minimizes J_k in Eq. (5.47). (Read the superscript 'o' as 'optimal'.) M_k is the sequence of bits in the time interval 0 to $k - L$ which minimizes J_k when the bits in the time interval $k - L + 1$ to k are prescribed; V_k is the corresponding minimum cost. (Both therefore depend on the sequence $(a_{k-L+1}, a_{k-L+2}, \ldots, a_k)$ prescribed.) Now V_k and M_k can be time-updated as follows:

$$V_{k+1}(a_{k-L+2}, a_{k-L+3}, \ldots, a_{k+1})$$
$$= \min_{a_0, a_1, \ldots, a_{k-L+1}} J_{k+1}(a_0, \ldots, a_{k+1})$$

$$= \min_{a_0, \ldots, a_{k-L+1}} \left[J_k(a_0, \ldots, a_k) + \left(\sum_{i=0}^{L} f_i a_{k+1-i} \right)^2 - 2\underline{Y}_{k+1} \sum_{i=0}^{L} f_i a_{k+1-i} \right]$$

$$= \min_{a_{k-L+1}} \left[V_k(a_{k-L+1}, \ldots, a_k) + \left(\sum_{i=0}^{L} f_i a_{k+1-i} \right)^2 - 2\underline{Y}_{k+1} \sum_{i=0}^{L} f_i a_{k+1-i} \right]$$

$$(5.49a)$$

and:

$$M_{k+1}(a_{k-L+2}, a_{k-L+3}, \ldots, a_{k+1})$$
$$= [M_k(a_{k-L+1}^o, a_{k-L+2}, \ldots, a_{k+1}), a_{k-L+1}^o] \quad (5.49b)$$

where a_{k-L+1}^o is the symbol which minimizes the right-hand side of Eq. (5.49a). The Viterbi algorithm propagates the 2^L values of $V_k(a_{k-L+1}, \ldots, a_k)$ together with the sequences $M_k(a_{k-L+1}, \ldots, a_k)$ from one value of k to the next according to Eqs. (5.49). Each of the 2^L sub-sequences $(a_{k-L+1}, \ldots, a_k)$ can be thought of as a possible state of the channel at time k. What comes out of the channel at time k depends on the last $L + 1$ bits transmitted, i.e., the channel 'remembers' the last $L + 1$ bits. For each of these states, $M_k(a_{k-L+1}, \ldots, a_k)$ is the most likely sequence of bits, $(a_0^o, \ldots, a_{k-L}^o)$, to have given rise to that state, and $V_k(a_{k-L+1}, \ldots, a_k)$ is the cost associated with that sequence. At the final time, N, the optimal sequence, $(a_0^*, \ldots, a_N^*)$, is made up of the state $(a_{N-L+1}^*, \ldots, a_N^*)$, which minimizes V_N, together with the sequence most likely to have given rise to that state, $M_N(a_{N-L+1}^*, \ldots, a_N^*)$, i.e.

$$(a_{N-L+1}^*, \ldots, a_N^*) = \arg \min_{a_{N-L+1}, \ldots, a_N} V_N(a_{N-L+1}, \ldots, a_N) \quad (5.50a)$$

and:

$$(a_0^*, \ldots, a_{N-L}^*) = M_N(a_{N-L+1}^*, \ldots, a_N^*) \quad (5.50b)$$

Because the Viterbi algorithm implements the MAP sequence criterion, the decision on the optimal sequence is deferred until the final time N. However, should the most likely sequences to have given rise to each of the 2^L possible states at time k coincide in the first m bits, then the decision on these bits can be taken immediately. Future observations $(\underline{Y}_{k+1}, \ldots, \underline{Y}_N)$ provide information about the state of the channel at time k, but whatever that state might be, the most likely subsequence of bits $(a_0^o, \ldots, a_{m-1}^o)$ to have given rise to it is the same. This so-called *merge* phenomenon can reduce the storage needed by hardware implementations of the Viterbi algorithm.

Even with merges, the Viterbi algorithm can be quite complex to implement, particularly if the extent of the i.s.i., L, is large. The algorithm needs to store at each time k the 2^L costs, V_k, and their associated most likely precursors, M_k. One way of reducing the complexity is to incorporate a linear equalizer before the implementation of the Viterbi algorithm in order to reduce the extent of the i.s.i. (The Viterbi algorithm is, of course, a *nonlinear* equalizer.) However, this recorrelates the noise sequence, resulting in nonoptimality of the Viterbi algorithm. As usual, design will be a compromise between strict optimality and low cost.

The MAP *symbol decision* criterion is developed by Abend and Fritchman (1970) and a Viterbi type algorithm for it is described by

Hayes *et al.* (1982); it is, however, somewhat more complex than the Viterbi algorithm for the sequence decision.

In this section it has been assumed that the bits transmitted over the channel are statistically independent. However, it is common practice with transmission systems to introduce redundancy into the data stream before transmission, as this assists error and fault diagnosis at the receiver. This subject will be studied in detail in Chapter 9. We note here that to make truly optimal use of the raw information arriving at the receiver with such systems, the process of *demodulation*, as described above, and *decoding*, as treated in Chapter 9, must be combined. In practice they rarely are, and the demodulator makes a so-called *hard decision* on the incoming bit stream, which it then passes on to the decoder. The decoder assesses independently whether or not the sequence contains errors and acts accordingly. This arrangement is nonoptimal because the decoder, on detecting an 'illegal' bit stream (i.e., one not consistent with the encoding strategy), is unable to make use of the unquantized information entering the demodulator when deciding which is the most likely 'legal' sequence to have been sent. Systems in which more information is passed from the demodulator to the decoder are often called *soft decision* systems as the demodulator makes only a 'soft' decision on the incoming bit stream and is prepared to change its decision appropriately in the event of a decision being deemed 'illegal' by the decoder.

5.6 Optimal detection of a frame alignment word

To make full use of high-capacity digital links it is common practice for data streams from a number of sources to be combined by *time division multiplexing*. This will be discussed in more depth in Chapter 7; here we look at the problem in the context of optimal decision-making. The idea is that symbols from each source are assembled into *frames* before being transmitted over a high-capacity link. At the receiving end of the link the symbols from each source must be extracted so that the original data streams can be reconstructed, and so the receiver of the multiplexed data must be able to recognize where the boundaries between frames lie. The usual technique is for the receiver to search the incoming stream of symbols for a predefined pattern called the *frame alignment word* (FAW), which is added to each frame by the transmitter. Since the data symbols in a frame are essentially random, and therefore able to imitate the FAW, care has to be exercised in the choice of a strategy for detecting its position.

Suppose that a frame comprises $n - m$ 'data' bits from the various

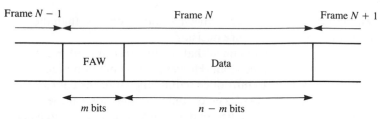

Figure 5.9   A typical frame structure

sources at the transmitting end together with an m-bit frame alignment word, and has the structure shown in Fig. 5.9. In any one frame period the receiver observes a sequence of n binary variables corrupted by independent zero-mean Gaussian variates with variance σ^2, i.e., the observations $(Y_0, \ldots, Y_{n-1})$ have the form:

$$Y_k = B_{k+a} + R_k \qquad (5.51)$$

where (R_k) is the sequence of Gaussian variates. The bipolar binary variables, B_i, are the bits that make up the frame:

$$B_i = \begin{cases} f_i & \text{if } 0 \leqslant i \leqslant m-1 \\ D_{i-m} & \text{otherwise} \end{cases} \qquad (5.52)$$

where $(f_0, \ldots, f_{m-1})$ is the frame alignment word and the D_i are the data bits. We assume that the data bits are statistically independent and that

$$\begin{aligned} P(D_i = -1) &= P(D_i = +1) \\ &= 1/2 \quad \text{for } 0 \leqslant i \leqslant n-m-1 \end{aligned} \qquad (5.53)$$

The additional term in the subscript of B in Eq. (5.51), a, which takes values in the set $\{0, 1, \ldots, n-1\}$, is the alignment parameter we wish to estimate. (All subscripts in this section are to be taken modulo n.)

If the prior probability that the frame boundary is in each of the n possible positions is $1/n$, then the MAP and ML decision criteria coincide; they pick the value of a which maximizes the density

$$p_{Y_0, Y_1, \ldots, Y_{n-1}|a}(Y_0, Y_1, \ldots, Y_{n-1}; a)$$

$$= \prod_{k=0}^{n-1} p_{Y_k|a}(Y_k, a)$$

$$= \prod_{k=0}^{m-1} n(f_k, \sigma^2)(Y_{k-a}) \prod_{k=m}^{n-1} \tfrac{1}{2}[n(-1, \sigma^2)(Y_{k-a}) + n(+1, \sigma^2)(Y_{k-a})]$$

$$(5.54)$$

where $n(\mu, \sigma^2)(y)$ is the Gaussian density with mean μ and variance σ^2 evaluated at the point y. Thus

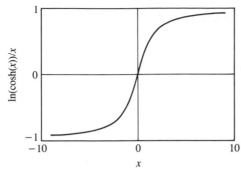

Figure 5.10 The function ln(cosh(x))/x

$$a^* = \arg\max_a \prod_{k=0}^{m-1} \exp\left(-(Y_{k-a} - f_k)^2/2\sigma^2\right)$$

$$\times \prod_{k=m}^{n-1} \tfrac{1}{2}\left[\exp\left(-(Y_{k-a} - 1)^2/2\sigma^2\right) + \exp\left(-(Y_{k-a} + 1)^2/2\sigma^2\right)\right]$$

$$= \arg\max_a \prod_{k=0}^{m-1} \exp\left(Y_{k-a}f_k/\sigma^2\right) \prod_{k=m}^{n-1} \cosh\left(Y_{k-a}/\sigma^2\right)$$

$$= \arg\max_a \left[\sum_{k=0}^{m-1} Y_{k-a}f_k + \sigma^2 \sum_{k=m}^{n-1} \ln\left(\cosh\left(Y_{k-a}/\sigma^2\right)\right)\right]$$

$$= \arg\max_a \sum_{k=0}^{m-1} \left[Y_{k-a}f_k - \sigma^2 \ln\left(\cosh\left(Y_{k-a}/\sigma^2\right)\right)\right]$$

$$= \arg\max_a \sum_{k=0}^{m-1} Y_{k-a}\left[f_k - \frac{\sigma^2}{Y_{k-a}} \ln\left(\cosh\left(Y_{k-a}/\sigma^2\right)\right)\right] \tag{5.55}$$

The function ln(cosh(x))/x is plotted in Fig. 5.10.

Thus the ML decision can be realized by a 'correlator' which scans through each m-bit sub-sequence of the observations, correlating it with an 'adjusted' form of the frame alignment sequence (the term f_k of the true correlator is replaced by $f_k - \sigma^2\ln(\cosh(Y_{k-a}/\sigma^2))/Y_{k-a}$), and then picks the sub-sequence showing the highest 'correlation'. By contrast, the strategy normally adopted in practice correlates each m-bit sub-sequence of the observations with the (nonadjusted) frame alignment sequence, and picks the sub-sequence showing the highest correlation. Strictly speaking, the ML strategy is not a correlator as the adjusted frame alignment sequence, with which the observations are 'correlated', is itself dependent on the observations. The adjustments have most effect when the signal-to-noise ratio of the system is high; they cause the 'correlator' to penalize observations, Y_{k-a}, which correlate negatively with their respective alignment sequence bits, f_k, more heavily that it rewards observations which correlate positively. This is intuitively exactly right: the probability that a true frame alignment bit is corrupted by noise is much smaller than the probability that a data bit, by chance, simulates a frame alignment bit.

The ML decision criterion given above was derived by Massey (1972). Massey went on to suggest an approximation to the optimal criterion which works well when the signal-to-noise ratio is high. Under such circumstances the variables Y_{k-a}/σ^2 are very likely to have large absolute values, which allows the approximation $\text{sgn}(x)$ to be used in place of the function $\ln(\cosh(x))/x$ in the adjustment terms.

In practical multiplex systems the frame alignment strategy may not have access to the continuous-valued samples from the output of the demodulator because sampling and quantization are usually performed in a single operation. Suppose, then, that the observations on which the alignment strategy has to operate are the bipolar binary random variables given by

$$Y_k = B_{k+a}R_k \tag{5.56}$$

where the R_k are independent bipolar binary random variables. The event $\{R_k = -1\}$ corresponds to a bit detection error at time k; consequently we set:

$$P(R_k = -1) = p < \tfrac{1}{2} \tag{5.57}$$

The ML criterion is now given by:

$$
\begin{aligned}
a^* &= \arg\max_a P(Y_0, Y_1, \ldots, Y_{n-1} \mid a) \\
&= \arg\max_a \prod_{k=0}^{m-1} [(1-p)(1 + Y_{k-a}f_k)/2 + p(1 - Y_{k-a}f_k)/2](\tfrac{1}{2})^{n-m} \\
&= \arg\max_a \prod_{k=0}^{m-1} [1 + (1-2p)Y_{k-a}f_k] \\
&= \arg\max_a \sum_{k=0}^{m-1} Y_{k-a}f_k \tag{5.58}
\end{aligned}
$$

which is the usual (nonadjusted) correlator. The probability of error for this criterion must clearly be greater than that for the ML criterion operating on unquantized samples; however it is simpler to implement. Whether it is worth implementing the more complex strategy or not depends on how great a reduction in probability of error can be achieved by so doing, and this depends on the signal-to-noise ratio, the frame length n, the frame alignment word, and its length m.

5.7 Summary

This chapter has described some optimal decision-making techniques and given examples of how they can be applied to digital transmission systems. As was stated in the introduction, the examples presented are illustrative and can be extended in a number of ways. For example, it is relatively easy to admit the cases of multilevel and complex-valued symbols (the latter occur in carrier-modulated systems) in the derivation of the optimal demodulator for a PAM signal. Also, the assumption that the Gaussian noise be white is unnecessary; it is used here as it simplifies the analysis a little. The Viterbi algorithm and related optimal symbol detectors are equally applicable to systems which use multilevel symbols and vectors of symbols; this includes complex-valued symbols and multisymbol words as special cases. For more details of these extensions and other applications of decision-making techniques refer to Cattermole (1986) and Proakis (1983).

References

Abend, K. and Fritchman, B. D. (1970) 'Statistical detection for communication channels with intersymbol interference', *Proc. IEEE*, vol. 58, no. 5, 779–85.

Cattermole, K. W. (1986) *Mathematical Foundations for Communication Engineering, vol. 2—Statistical Analysis and Finite Structures*, Pentech, London.

Forney, G. D. (1972) 'Maximum-likelihood sequence estimation of digital sequences in the presence of inter-symbol interference', *IEEE Trans. Inform. Theory*, vol. 18, no. 3, 363–378.

Hayes, J. F., Cover, T. M. and Riera, J. B. (1982) 'Optimal sequence detection and optimal symbol-by-symbol detection: similar algorithms', *IEEE Trans. Comm.*, vol. 30, no. 1, 152–7.

Massey, J. L. (1972) 'Optimum frame synchronization', *IEEE Trans. Comm.*, vol. 20, no. 2, 115–19.

Papoulis, A. (1972) *Probability, Random Variables and Stochastic Processes*, McGraw-Hill, New York.

Proakis, J. G. (1983) *Digital Communications*, McGraw-Hill, New York.

Viterbi, A. J. (1967) 'Error bounds for convolutional codes and an asymptotically optimum decoding algorithm', *IEEE Trans. Inform. Theory*, vol. 13, 260–9.

PART 2

Signal multiplexing and coding

6 Digital multiplexing principles

STEVE WHITT

6.1 Multiplexer or concentrator?

Both multiplexers and concentrators aim to convey information or data
from many independent sources over a much reduced number of
transmission links. Usually, the many sources are geographically
colocated and wish to communicate with some distant point in a
network, but if each source has an individual physical link travelling a
long distance across the network, the overall cost of the network is
prohibitive. However, this is where the similarity between concentrators
and multiplexers ends.

Concentration takes advantage of the situation where the separate
inputs into a system are not all simultaneously active. This is typical of
most telephone or data networks where terminals are active on average
for under 20 per cent of the time. A simple example is the situation
where a group of computer terminals located in one building needs to
access a central host computer located in a distant town. It is clear that
a one-to-one match between the number of terminals and the number of
connections to the host computer is not necessary; if the terminal usage
was a constant percentage of the time this would define the ratio of
connections to terminals but, in practice, terminal usage is a stochastic
process. Operationally solving the relationship between terminal usage
and connection dimensioning is fundamental to cost-effective design of
data and telephone networks and is the *raison d'être* of teletraffic
engineering.

Multiplexing, on the other hand, is a means of sharing a
communications link that does not employ the contention of a
concentrator thus avoiding possible queueing and delay in busy periods.

The easiest way to distinguish between a concentrator and a
multiplexer is to examine its data throughput ratio, defined as:

$$\frac{\text{Sum of all input data rates}}{\text{System output data rate}}$$

For a concentrator this ratio is greater than 1 because not all the input data can appear simultaneously at the output, whereas for a multiplexer it is less than 1 because even if all tributaries are active no data is lost in the multiplexed form appearing at the output. The ratio tends towards unity for a poorly utilized concentrator and for a well-utilized multiplexer. The main purpose of multiplexers is to reduce data transmission costs by using a single physical bearer to carry a number of simultaneous transmission circuits instead of using one bearer per circuit. Multiplexing can employ a range of techniques, of which *frequency division multiplexing* (FDM) and *time division multiplexing* (TDM) are the most widely used, the latter being the prime means of digital multiplexing on which the rest of this chapter will focus.

6.2 TDM versus FDM

Although this chapter deals with TDM principles it is worth while recognizing the reasons why FDM emerged first, only to be superseded by TDM in more recent years.

Historically, FDM emerged as the predominant multiplexing technique for two reasons. Firstly, it was an analogue technique and this was clearly compatible with analogue information (e.g., speech, music) that needed to be conveyed. Secondly, the electronic circuits and devices available were not known for their high speed or great bandwidth and this meant economy of bandwidth was essential. For example, an analogue speech circuit needs less than 4 kHz of channel bandwidth, whereas a simple digital representation of the same speech (without sophisticated digital signal processing) is transmitted at 64 kbit/s and needs at least 32 kHz channel bandwidth.

Until the widespread emergence of the computer in the 1960s there had not been any need to consider the transmission of information that originated in a digital form (apart from telex). The progress of digital computers mirrored the major advances in electronic technology and circuit design which, by the 1970s, were beginning to offer the potential of very high transmission bandwidth and very large scale integrated circuits, both of which are facilitators for digital transmission and TDM. In the major economic nations of the world digital transmission and TDM, together with digital switching, have been deployed since the late 1970s, but introduction has been a slow process due to the pre-existence of a complete analogue telecommunications network which had to be modernized and replaced at considerable cost. By the 1990s, analogue networks were becoming confined to the local loop portion of the network (i.e., that portion between customers' premises and the nearest building belonging to the telephone company). In British Telecom's network, for example, since the summer of 1990 the entire trunk network used for switched telephony has been digital, employing TDM throughout.

The key benefits of TDM over FDM are:

- channel-to-channel performance uniformity,
- ease of cascading many multiplexers and more flexible design,
- compatibility with digital switching,
- compatibility with digital traffic sources (e.g., computers),
- digital techniques allow more sophisticated performance monitoring of equipment and service quality, e.g., parity checks and error correction,
- digital signals and circuits are more amenable to large scale circuit integration, leading to space, power, and cost savings.

6.3 Multiplexing economics

Whether or not it is advantageous to use multiplexers to aggregate a number of low-speed channels into a single higher speed channel is dependent primarily on the cost of transmission versus the cost of multiplexing. To illustrate the principle it is worth considering a simple problem often faced in long-distance (trunk or toll) transmission networks operated by telephone companies. The case illustrated in Fig. 6.1 needs to convey four 100 Mbit/s channels over a certain distance D. The planner can choose to transmit four separate 100 Mbit/s signals on individual transmission systems (Fig. 6.1a) or he can employ a single 400 Mbit/s transmission system with a 4:1 multiplexer at one end and a 1:4 demultiplexer at the other (Fig. 6.1b).

Using the hypothetical prices shown in the figure, which allow for the fact that equipment operating at 400 Mbit/s costs more than its 100 Mbit/s equivalent, both options can be costed to give:

$$\text{Cost of multiplexed option} = £26\,000 + £1500D$$
$$\text{Cost of nonmultiplexed option} = £16\,000 + £4000D$$

Examining the route distance D as the variable, these equations tell the planner that both options cost the same when $D = 4\,\text{km}$. It can also be seen that longer routes are cheaper if multiplexers are used, whereas short routes would be better built without multiplexers.

The above example assumes that four data channels needed to be conveyed, but would the solution be the same if only two channels were operational? Assuming the same 4:1 multiplexing equipment is used, albeit with only 50 per cent utilization, the equality is:

$$£26\,000 + £1500D = £8000 + £2000D$$

which gives the breakeven distance as 36 km. This simple example illustrates that on long-haul routes it is generally economic to multiplex

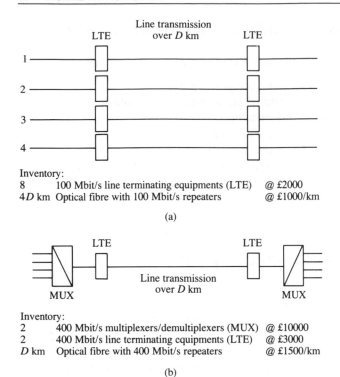

Inventory:
8   100 Mbit/s line terminating equipments (LTE)   @ £2000
4D km Optical fibre with 100 Mbit/s repeaters   @ £1000/km

(a)

Inventory:
2   400 Mbit/s multiplexers/demultiplexers (MUX)   @ £10000
2   400 Mbit/s line terminating equipments (LTE)   @ £3000
D km   Optical fibre with 400 Mbit/s repeaters   @ £1500/km

(b)

Figure 6.1   Basic multiplexing economics

data up to a higher level than may be immediately necessary, in which case the route is provided to cope with future growth in traffic. If near future growth in demand can be anticipated then the equation becomes complicated by factors such as rate of return on capital, and the benefits of deferred capital investment. However, if the above example is modified such that the initial system utilization of 50 per cent is augmented by new demand so that the multiplexer becomes 75 per cent utilized after two years' use, then the breakeven distance for the multiplexed option will be around 10 km.

The earlier example assumed a multiplexing ratio of 4:1, but another key economic question is: what is the most economic multiplexing ratio? A multiplexer with the smallest ratio of 2:1 is generally easy and cheap to design, but only a limited transmission advantage is obtained. At high multiplexing ratios one often is limited by technology capabilities and hence escalating multiplexer cost. Additionally, before monomode optical fibre was commonly deployed in telecommunications networks, available bandwidth on copper cable and radio systems was often the factor limiting the multiplexer output data rates. Historically, the multiplexing ratio has been almost universally around 4:1 for systems operating above about 1 Mbit/s. Indeed, multistage multiplexing has

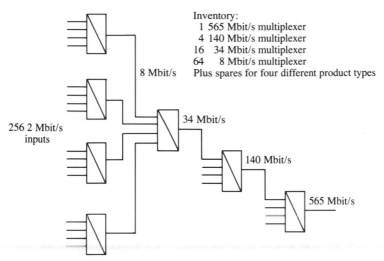

Figure 6.2   The CCITT European multiplex mountain

usually been achieved by cascading several separate 4 : 1 multiplexers; this in general would not be as economic as a single multiplexer with a large multiplexing ratio, were the equipments to be purchased new, but the inefficient *multiplex mountain* (Fig. 6.2) has gradually evolved in stages from low speed to high speed, as demand for transmission capacity has grown over the years and as newer technology has enabled electronic circuits to operate at higher data rates.

Although there are clear economic advantages associated with higher degrees of multiplexing, these attractions need to be balanced against the dangers of 'having too many eggs in one basket'. Failure of the transmission path, or the multiplex equipment, is inevitable and will result in complete loss of all communications being carried. The network designer needs to consider the impact of such failure and the means of reducing the risk of its occurrence, methods for restoring service, and the consequences for the basic cost of providing service.

6.4 Aspects of digital multiplexing

The basis of multichannel digital communications systems is the time division multiplexer, which works on the principle of taking a sample of data from a number of sources, known as tributaries, and allocating each of these a period of time, known as a timeslot, in which onward transmission occurs. Individual timeslots are assembled into frames to form a single continuous high-speed data stream. A frame is the smallest group of bits in the output data stream containing at least one sample from each input channel plus various overhead information that will

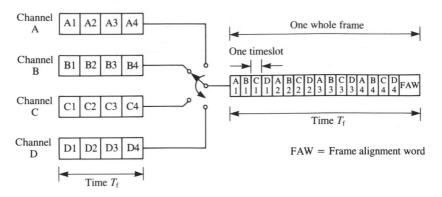

Figure 6.3 Basic bit-interleaved TDM

allow the individual timeslots to be identified and subsequently separated back into individual channels.

6.4.1 Bit-interleaved TDM

If a multiplexer assigns each incoming channel a timeslot equal to one bit period of the outgoing data stream, the arrangement is known as *bit interleaving* (Fig. 6.3). In this case the multiplexer can be visualized as a commutator switching sequentially from one input channel to the next. In principle this is a simple means of realizing a TDM multiplexer, but it can be seen that all the electronic circuitry is operating at the speed of the output data stream and this can be a problem in very high capacity multiplexers which may have output data streams in the order of Gbit/s. The high-speed circuitry generally increases the cost and power consumption and reduces the scale of practical circuit integration. Therefore, for very high capacity multiplexers, deliberate attempts are made to use as much parallel processing as possible to minimize the quantity of electronics working at the output data rate. It is often assumed that all tributaries are operating at the same low data rate before interleaving to produce the higher speed output rate, but this need not be so as long as different inputs have a fixed relationship between their data rates (Fig. 6.4). In practice, multiplexers with mixed input rates have been reserved usually for low-speed applications (i.e., sub 2 Mbit/s), but in the near future high-speed multiplexers with outputs greater than 100 Mbit/s are likely to be capable of handling mixed inputs (Newall, 1992).

6.4.2 Byte- or word-interleaved TDM

Instead of sampling each incoming channel on a bit-by-bit basis, it is possible for a multiplexer to accept a group of bits, making up a word or a character, from each channel in turn. In the predominantly binary

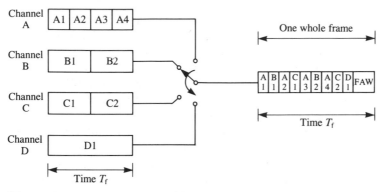

Figure 6.4   Bit-interleaved TDM with differing input data rates

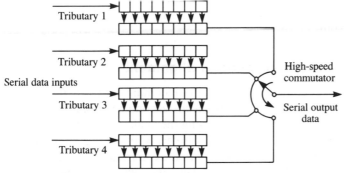

Figure 6.5   Simplified byte-interleaved time division multiplexer

digital environment it is practical to handle blocks of 2^n bits, with $n = 3$ being particularly popular, hence the widespread use of byte interleaving.

This scheme introduces the need for buffer storage of data incoming to a multiplexer, since bits need to be accumulated on each input before being interleaved to form the output data stream. A simplified byte-interleaved multiplexer is shown in Fig. 6.5, in which incoming data is accumulated at low speed in the buffer store. When a complete word has been received it is transferred in parallel to the register and read out at high speed to form a byte-sized timeslot in the outgoing data stream.

6.4.3 Synchronicity

So far in the simple examples illustrated we have assumed that the incoming data streams are synchronous with each other and with the internal timing of the multiplexer. If there is any difference in the timing of the input data relative to that of the multiplexer, additional signal processing is necessary which adds to the design complexity of both multiplexer and demultiplexer.

If the digital sources are nearly synchronous with the multiplexer and any difference constrained within predefined frequency limits (known as *plesiochronous* sources), then a technique known as justification or bit-stuffing is commonly used to compensate for the small frequency differences. In practice, these differences can be due to static offsets caused by drift of clock frequency, or dynamic changes resulting from clock jitter or wander associated with the incoming data.

Synchronous operation is clearly desirable since it simplifies the design (and reduces the cost) of multiplexers but, in practice, it remains difficult to maintain true network synchronicity over wide geographical areas. As a consequence only multiplexers that are likely to have inputs from a small geographical area have been designed for synchronous operation, for example 30-channel 2 Mbit/s multiplex equipment described in detail in CCITT G736 in which the 64 kbit/s inputs are synchronous with the 2048 kbit/s output.

6.4.4 Frame synchronization or alignment

When a multiplexer combines a number of incoming data streams into an aggregate output, it is essential that a corresponding demultiplexer can identify which channel is which. Using the analogy of the commutator switch referred to earlier, it is necessary for the commutator in the demultiplexer to remain in step with the commutator in the multiplexer, otherwise the demultiplexed channels will be presented to the wrong output ports.

The usual means of aligning the demultiplexer with incoming data is a frame alignment word, which is a specific code interleaved and transmitted with the data leaving the multiplexer. This code is recognized by the demultiplexer, which has inherent knowledge of where in the frame such a code should appear. The demultiplexer therefore can make its own standalone decisions regarding frame alignment, and this forward-acting frame alignment method provides for automatic recognition and recovery from loss of frame alignment. The frames illustrated in Figs. 6.3 and 6.4 both include frame alignment words for this purpose.

Some multiplexers (primarily low speed) use a frame synchronization technique known as *handshaking*, since it involves interactive cooperative working between the sending multiplexer and the receiving demultiplexer. Usually handshaking takes the form of a synchronizing preamble transmitted prior to the main message, which has the advantage of simplicity and channel bandwidth efficiency, but if long messages are to be transmitted without loss of synchronization the transmission channel needs to be error- and break-free.

The analysis of frame realignment, loss of alignment, frame mimicking, and error extension are complex subjects and will not be covered further here since frame alignment is discussed in some detail in Chapter 7 and error extension is covered in Chapter 8.

6.4.5 Justification or bit-stuffing

As mentioned previously, a multiplexer intended to handle plesiochronous data inputs needs to employ a technique known as justification or bit stuffing, in which the input data is converted into a truly synchronous signal suitable for interleaving into an outgoing frame structure. Bit-stuffing, as its name implies, adds (or in certain cases subtracts) extra bits to (from) the incoming data to create a synchronous stream of data. If the incoming rate varies because of drift of a clock frequency or jitter then the degree of stuffing that takes place needs to vary dynamically. In order that the demultiplexer can reconstruct the original data streams it needs to know when stuffing has occurred and which bits of data are stuff bits, so that they can be selectively removed and discarded. This implies some form of communication between multiplexer and demultiplexer which commonly takes the form of justification control bits included at a preassigned position in the frame generated by the multiplexer. There are several approaches to justification, each with their merits. Some are simpler to implement, others create less waiting time jitter or less pulse stuffing jitter and so on. Well-known techniques include positive justification, positive–negative justification and positive–zero–negative justification (Smith, 1985).

With positive justification, for example, each tributary rate is synchronized to a rate slightly greater than the maximum allowable input rate. For instance, in the plesiochronous digital hierarchy (discussed in Sec. 6.5.1) a nominal 140 Mbit/s data rate is actually 139 264 kbit/s ± 15 p.p.m. Positive justification needs to stuff enough bits to increase the incoming rate to a new synchronous rate above 139 267 kbit/s, which is the maximum rate allowed for a nominal 140 Mbit/s data stream. If the input rate is running fast, fewer stuff bits are added than if the input is running slowly and, as long as the input data rate stays within its design limits, the sum of data bits and stuff bits is kept constant. In many ways the demultiplexing process is more complicated since not only do the stuff bits need to be removed, but a new clock also needs to be generated to time each demultiplexed tributary. Since each tributary can be instantaneously operating at any rate within the plesiochronous limits, the demultiplexer needs a separate clock per tributary to recreate the original timing, which can be quite a serious practical design burden.

One of the most serious problems with justification is the generation of substantial amounts of jitter, which will influence the choice of justification scheme (Smith, 1985) or encourage the use of synchronous multiplexing. Additionally, since stuff bits and justification control information affect the whole demultiplexing process, any errors introduced into these parts of the frame will result in serious error multiplication unless specific precautions, discussed in Chapter 8, are incorporated.

6.4.6 Statistical multiplexing

Although it seems that there is a clear distinction between a concentrator and a multiplexer, there is a family of multiplexers that would appear to be a hybrid concentrator/multiplexer. This family of statistical multiplexers is usually encountered as equipment employing digital speech interpolation (DSI). DSI is a technique that exploits the fact that in telephone conversations neither speaker is speaking all the time and that even in continuous speech there are significant gaps between words and syllables. In a statistical multiplexer each channel is only connected to the multiplexer input during periods of activity, allowing more channels to be handled than by conventional TDM techniques alone. Studies by Chu (1969, 1971) have shown that statistical TDM should be able to handle between two and four times as many speech channels as a conventional TDM system. This technique obviously has advantages, but there are attendant problems. More storage is needed at the multiplexer to hold incoming messages and at the demultiplexer to wait for the arrival of information before reassembly into individual channels. There is a risk of excessive delays due to channel congestion and there is the penalty of more complex message assembly, system control, and management. This delay may be sufficient to warrant the use of echo suppression on speech circuits and if these do not pre-exist on the path, the cost of introducing them needs to be assigned to the multiplexer. It must be remembered that a multiplexer is used only if the savings in transmission costs exceed the multiplexer cost. For this reason, complex and expensive multiplexers such as these will only be justified on economic grounds on long-distance lines, or in special cases where there is a finite limit on bandwidth such as on certain satellite circuits. In practice, DSI is widely used on international telephone circuits where multiplication of ratios 3–5 are achieved.

6.5 Digital multiplexing hierarchies

Since the initial development of digital telecommunications networks in the 1970s several different de facto standards have emerged. This has resulted in a legacy of three different multiplexing hierarchies adopted in different parts of the world. The deliberate acceptance of regionally unique multiplexing hierarchies was, in part, influenced by technical

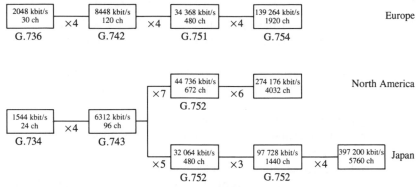

Figure 6.6 The plesiochronous digital hierarchy. (The numbers in the boxes represent the output data rate of each multiplexer and the number of 64 kbit/s channels carried. The reference below the box indicates a CCITT recommendation relevant to that multiplexer.)

considerations, but market protection and the cosy relationship between domestic equipment manufacturers and their monopolistic telephone company customers held just as much sway in the decision-making process. However, in a shrinking world in which international communications is becoming ever more important, it is vital to have agreed standards governing the interconnection of networks from one nation to another. To this end the CCITT has adopted and regulated the hierarchies, which historically reflect the choices made by AT&T in the USA, NTT in Japan and the CEPT in Europe. Unfortunately, the only level common to all three is the 64 kbit/s base level. Until recently, these three largely incompatible hierarchies were the only ones in existence but since 1985 a new *synchronous digital hierarchy* (SDH) has been developed to overcome the technical limitations of the established *plesiochronous digital hierarchy* (PDH) and an attempt has been made to overcome the problems associated with international incompatibility.

6.5.1 Plesiochronous digital hierarchy

The first level of each CCITT digital hierarchy is achieved by the synchronous multiplexing of a number of 64 kbit/s channels. This produces a bit rate of 1544 kbit/s (24 channels) in North America and Japan but 2048 kbit/s (30 channels) in Europe. Figure 6.6 shows the structure above these rates and clearly shows the incompatibility between systems. Although each multiplexing step achieves an integer multiplexing ratio, the bit rate increase is somewhat greater than the increase in channels carried. For example, in the North American hierarchy a 1544 kbit/s system carries 24 × 64 kbit/s PCM channels, but on subsequent multiplexing with three other 1544 kbit/s data streams the second-order multiplexer produces a data stream with an output data rate of 6312 kbit/s which is greater than precisely 4×1544 ($= 6176$ kbit/s). The output still only carries $4 \times 24 = 96$ PCM channels, but a degree of transport inefficiency has to be accepted since some overhead data is inserted in the frame to ensure frame alignment and to cope with the plesiochronous nature of the incoming 1544 kbit/s data.

Although the highest standardized bit rate is 397 200 kbit/s, multiplexers with nominal output data rates of 565 Mbit/s are widely deployed in the USA and Europe. Some of these newer multiplexers employ sophisticated techniques and technology in an attempt to overcome some of the limitations of the PDH (Brooks *et al.*, 1985; 1986) but in overall network terms these efforts remain largely cosmetic in nature.

Full technical definitions of all three variants of the PDH can be found in recommendations published by the CCITT as recommendations G731–755.

The main drawbacks of the existing PDH can be summarized as:

- lack of international standards above 140 Mbit/s and incompatible standards below,
- lack of design flexibility restricts range of products and potential of telecommunications networks,
- need to concatenate equipments to build a multiplexer mountain which is costly and unreliable,
- interface standards based on electrical signals on copper become a serious problem above 140 Mbit/s because of limited transmission distances over coaxial cable,
- lack of embedded standard operation and maintenance facilities,
- low data throughput ratios for the complete multiplex mountain because of poor practical equipment utilization (although the ratio in operational equipment may be on average 80 per cent for each multiplexer, when four are cascaded the overall figure drops to just $0.8^4 \times 100 = 41$ per cent).

Although it would theoretically be possible to overcome these limitations by designing a totally new generation of PDH, it makes sense to exploit such an opportunity to grasp the additional advantages of a synchronous approach to multiplexing and network design.

6.5.2 Synchronous digital hierarchy

Before going into the detail of the synchronous digital hierarchy it is worth briefly examining its origins, since this reflects the importance of having a workable international standard for something so fundamentally important to telecommunications. Attempts to produce a set of standards covering optical transmission of synchronous signals first started in the USA in early 1984 in response to a particular problem being faced by regional telephone companies. The problems of sourcing optical products from more than one supplier and their successful interconnection led to the concept of the standardized optical 'midspan meet'. Considerable work had also been undertaken in the USA on synchronous transmission systems that proposed a standard known as Syntran, defining a system that multiplexed 64 kbit/s channels into a 45 Mbit/s base transmission rate. In parallel, AT&T, also in the USA, was developing a proprietary synchronous system known as Metrobus, based on a 150 Mbit/s transmission rate.

The Americans were not alone in developing synchronous systems and in July 1984 British Telecom started the development of a synchronous multiplexer, the S Mux 1401, that combined 64 × 2 Mbit/s channels into a 140 Mbit/s transmission rate (Hall and McCartney, 1987). Some of these early proprietary systems went into production and the S-Mux 1401 was operationally deployed in the British Telecom junction network. However, it was soon recognized that international standards would be needed to cover what were becoming very sophisticated pieces of multiplex and transmission equipment.

In 1985 the T1X1 committee of the American National Standards Institute (ANSI) decided to create standards to address the complete issue of synchronous networks as proposed originally by Bellcore in the USA. Hence Synchronous Optical Networks (SONET) was born, described by Ballart and Ching (1989). As a result of much work on proprietary systems there were many contentious issues to be resolved before standards could be agreed. To complicate matters, European telephone administrations had shown little interest in SONET, but it became apparent that if nothing were done to internationalize SONET the world would continue to have different multiplexing hierarchies in different areas, even after the move to synchronous networks. Initially British Telecom (shortly followed by Telecom Australia and Televerket in Sweden) intervened to ensure that any new standards would be suitable for the 2 Mbit/s areas of the world as well as the 1.5 Mbit/s regions.

In the summer of 1986 the CCITT took an interest in SONET and began work on the synchronous digital hierarchy. Subsequently, there were a couple of false dawns as the multiplexing frame structure went through major redesign, and there was considerable commercial pressure to finalize standards quickly. After considerable debate worldwide on what was an extremely complex topic, encompassing not just multiplexing but optical transmission, network operations, equipment supervision, and software control, agreement between the CCITT and the T1X1 committee of ANSI was reached in 1988.

The CCITT produced three recommendations to cover the SDH—

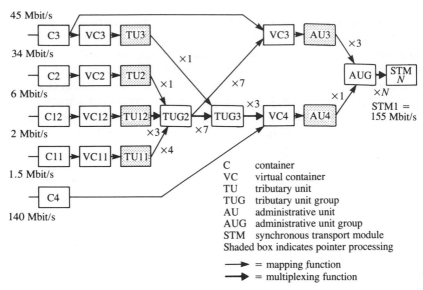

Figure 6.7 The synchronous digital hierarchy

G.707, G.708 and G.709—which describe the subject in considerable detail.

SDH introduced a number of new concepts in multiplexing and, because the standards have been designed by committee to cope with both the 2 Mbit/s and the 1.5 Mbit/s plesiochronous hierarchies, there is inevitably considerable complexity. Indeed, it can be argued that simplicity of frame structure has been sacrificed to achieve a higher degree of flexibility than most users would need, the consequence of which is more complex equipment designs and somewhat higher costs.

Figure 6.7 shows the range of options available in the SDH, but to illustrate the processes involved it is worth examining how an input data stream (on the left) eventually appears as part of the multiplexed data stream. Starting with a conventional plesiochronous input data stream, an SDH equipment initially applies justification or bit-stuffing to it to form a synchronous signal. The data plus justification remain together in a container (C) until the data finally leaves the synchronous network. To ensure a high level of performance monitoring on a per channel basis, special information known as *path overhead* is added to each container to create a virtual container (VC). Again these path overheads remain with the container until the data leaves the network. One of the problems with synchronous networks is the difficulty of ensuring true synchronicity across large geographical areas so, as a virtual container moves through a network or from one network to another, its aggregate data rate may need adjustment by applying a degree of justification. After the necessary justification the virtual container becomes a tributary

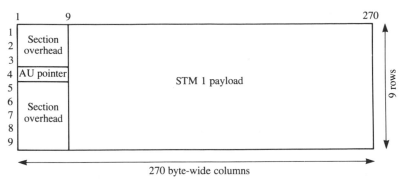

Figure 6.8   The STM1 simplified frame structure

unit (TU). Because the VC is not necessarily truly synchronous with the
TU its exact position relative to the TU is marked with a pointer, which
is additional information also carried within the TU. This facilitates
easy location of each VC within the aggregate frame structure. TUs can
be simply grouped together to form a tributary unit group (TUG)
before being mapped into a larger VC.

The process of loading containers and attaching overheads is repeated
at several levels in the SDH, resulting in the nesting of small VCs
within larger ones. The process is repeated until the largest VC (the
VC4 in Europe) is full and this is then loaded via the administrative
unit (AU) into the payload of the synchronous transport module
(STM). AUs are logically combined into different size groups (AUGs)
according to the intended transmission rate; e.g., three AU3s are
mapped into an STM1 operating at a line rate of 155.52 Mbit/s. At this
point a synchronous data stream exists with a frame containing payload
and section overheads carefully arranged to ensure simple and direct
access to information contained within. Section overheads are bytes
reserved for communication between adjacent pieces of synchronous
equipment (e.g., multiplexers and regenerators). As well as being used
for frame synchronization, they perform a variety of management and
administration functions. The STM1 frame has a 125 μs repetition rate
and is built up of nine rows each of 270 bytes (i.e., frame length is
19 440 bits) that can be stacked for clarity of visual representation (Fig.
6.8). Data is simply transmitted in sequence from left to right taking
each row in sequence, starting out at the top left-hand corner of the
diagram. Since all signals at STM1 are synchronous, it is very simple
to create higher transmission rates by byte-interleaving payloads
without any additonal complexity of justification or additional
overheads.

Although there are obviously many permutations possible in the
multiplexing hierarchy, a few have been chosen by standards
organizations as their preferred options. For example, the European

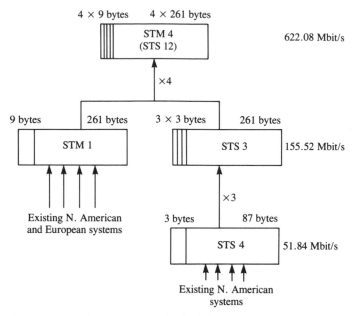

Figure 6.9 Comparison of SONET and SDH signals

Telecommunications Standards Institute (ETSI) has selected the following route for carrying 2 Mbit/s data:

Input: C12 → VC12 → TU12 → TUG2 →
$$\qquad\qquad\qquad\qquad TUG3 → VC4 → AU4 → STM1$$

In significant contrast, ANSI has opted for an SDH subset that reflects its earlier origins in SONET. Both the ANSI and ETSI standards select the TUG2, but from here they diverge: while Europe opts for the AU4 path, North America has chosen the AU3 path, resulting in a 51.84 Mbit/s signal (called an STS1 in SONET terminology). However, three STS1 signals multiplexed together (i.e., an STS3 signal) assume a frame format entirely compatible with the STM1 format and thus full interworking is achieved at all rates from 155 Mbit/s upwards (Fig. 6.9).

Although multiplexing in SDH is a multistage process, the boxes in Fig. 6.7 do not represent individual pieces of equipment (as is the case in the PDH shown in Fig. 6.6), rather they represent logical functions undertaken within one unit.

Although the original thoughts of designers working on Metrobus, Syntran and the S-Mux 1401 focused on a truly synchronous network and the simplicity of synchronous multiplexer design, the SDH is now capable of handling as an input virtually any bit rate used in the existing PDH. In fact, these inputs need not be synchronous at all, since justification is specified for input tributaries. Some of the new concepts introduced by SDH, along with a completely new nomenclature, cannot be covered in sufficient detail here and further reading is recommended (Newall, 1992; IEEE Communications, 1990).

It is, however, worth summarizing the key benefits that SDH brings in order to justify the trauma of its birth and forthcoming introduction into networks in the mid-1990s:

1   A very flexible frame structure permits transmission and multiplexing of a wide variety of data types (everything from 64 kbit/s to 140 Mbit/s). It was designed, as far as possible, to be future-proof, in that SDH forms the physical layer for the broadband ISDN, and SDH has been designed to be compatible with the emerging asynchronous transfer mode (ATM) of data transmission (de Prycker, 1991).

2   Substantial monitoring and control facilities included as overheads within the frame structure enable the design of intelligent, flexible, and controllable synchronous multiplexers to replace their dumb equivalents operating in the PDH. This opens the door to advanced, remotely controlled, network management and genuine end-to-end performance monitoring of customer circuits through the network, irrespective of their routeing.

3   Greater multiplexing flexibility will facilitate efficient and reliable design of add–drop multiplexers and generate the potential for new automated higher order cross-connects. (Cross-connects are the electronic equivalent of digital distribution frames or coaxial cable patch panels used widely today as network flexibility points.) New multiplexer types pave the way for new and more flexible network architectures, which will ease the introduction of new services and offer improved path availability (i.e., reduced down time as perceived by the customer).

4   International standards will lead to greater competition in equipment supply and lower costs, leading to considerable cross-border sales which were absent in the PDH telecommunications market.

6.6 Multiplexing network architecture

Since its earliest days, telephony has primarily concerned itself with point-to-point communications so, as networks developed in sophistication, there remained a tendency to view customer-to-customer links as a concatenation of numerous point-to-point communications systems. As a consequence, multiplexing techniques (both FDM and TDM) centred on very simple linear network structures using two terminal multiplexers and some form of linking transmission system.

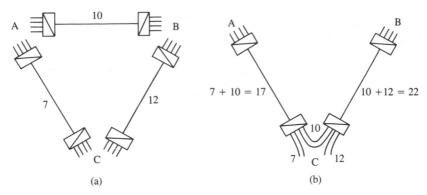

Figure 6.10 Network architectures using terminal multiplexers

There are, however, two problems inherent in this simple architecture. Firstly, a point-to-point system is vulnerable to any failure of its component parts, leading to interruption of service. Either the fault needs to be repaired in real time, which can still lead to lengthy interruption, or the traffic needs to be switched to a secondary or standby path which enables communication while the first fault is repaired. In the latter case, spare equipment must be provided for such an eventuality, together with a means of protection switching from main to standby paths.

The second problem with simple point-to-point multiplexed transmission links is a question of economics; not whether it is economic to multiplex or not, but the wider network-level economics. Imagine a simple network made up of three towns, each initially linked by point-to-point systems carrying the number of traffic channels shown in Fig. 6.10(a). Assume that the number of traffic channels shown justifies the installation, but does not fully utilize a multiplex line system on each leg AC, AB, and BC when analysed in isolation. One might assume that this was a cost-effective network, unless the wider perspective is addressed; this simple network needs three transmission systems and six multiplexers. Still using conventional multiplexers it is possible to create a more economic network in which traffic from A to B is routed via C (Fig. 6.10b). However, this construction impairs the reliability of the through traffic path from A to B since it passes through twice as much failure-prone equipment as in the initial scheme (Fig. 6.10a). On the positive side, though, the cost is reduced by a third since only four multiplexers and two transmission systems are needed.

There is, in fact, a still cheaper and more reliable way of building the network that relies on what is known as a drop-and-insert multiplexer or *add/drop multiplexer* (ADM). Whereas a conventional terminal multiplexer is a two-port device (tributary and line) with only two possible directions of traffic flow, an ADM has three ports (tributary,

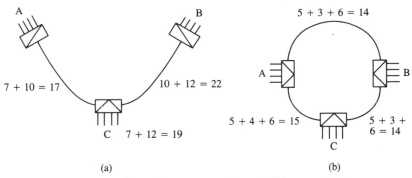

(a)   (b)

Figure 6.11   Network architectures using add/drop multiplexers

line 1 and line 2) and six possible traffic flows. This increase in potential connectivity within the multiplexer opens the door to cheaper and more flexible networks.

With this new form of multiplexer it is possible to construct a solution based on just two transmission links, two terminal multiplexers and one ADM (Fig. 6.11a). This solution uses the least equipment of all and can overcome the reliability problem of concatenated equipment to some extent by relying on an appropriately reliable design of ADM. Indeed, this is the key reason why ADMs have not proliferated in networks in the past, since in a plesiochronous environment the design is extremely complex and costly, and through path traffic reliability would be poor. However, in the last few years, with the advent of synchronous higher order multiplexing, a new generation of ADMs has been born. In fact, the first commercially produced ADM designed with full ring facilities in mind was deployed by British Telecom in the UK in 1989 using a proprietary synchronous multiplexing method developed in advance of SDH standards (described by Hall and McCartney (1987)).

Although Fig. 6.11(a) illustrates a combination of ADMs and terminal multiplexers working together, it is often advantageous to build a network purely of ADMs and to create a *ring* (Fig. 6.11b). This ring can provide the full functionality required by the network in Fig. 6.10(a), and much more.

The first thing that should be apparent in the structure shown in Fig. 6.11(b) is the presence of two independent routes between any pair of nodes; one clockwise and one anticlockwise. This immediately offers greater path availability, since if a fault breaks the clockwise route between any two nodes, then use of the corresponding anticlockwise route will ensure minimum service interruption. This is just one example of a whole range of protection strategies that have been developed for ADM rings in recent years, discussed by Whitt *et al.* (1990) and Helmes (1989). Figure 6.11(b) illustrates a scheme that further enhances network resilience by sending half the traffic clockwise and the remainder anticlockwise around the ring. Another particularly attractive scheme avoids the need to switch the transmission path from clockwise to anticlockwise since it actually broadcasts simultaneously in both directions. The receiving node decides which of the two notionally identical sets of data it is receiving is better quality and proceeds to select that signal. This particular scheme is attractive because it eliminates the need for node-to-node command and control communications under fault conditions, but it is inevitably wasteful of transmission capacity.

Rings have an additional feature that makes them particularly attractive in a network trying to cope with unpredictable localized growth in demand for transport capacity (i.e., the real world). Returning to Fig. 6.10(a), consider the situation where there is a growth in demand between A and B from the 10 channels to, say, 30 channels. If this exceeds the carrying capacity of the existing system AB then a new system AB2 would have to be installed, even though there may still be plenty of spare capacity on route AC and CB. The inflexibility of the conventional linear multiplex structure makes accessing any such spare capacity rather difficult. In contrast, the ring structure shown in Fig. 6.11(b) would be able to set up transmission quickly to carry the extra capacity, some clockwise and some anticlockwise, without reaching capacity saturation in either direction. It is the simple fact that an ADM has two ways out for its input tributary traffic, that can afford the whole network additional flexibility and resilience. In operational situations, such flexibility is most valuable on rings made up of many nodes that normally would not generate enough traffic really to justify full direct node-to-node interconnection.

6.7 The future

This chapter does not conclude the story of digital multiplexing and, indeed, in the space available it has only been possible to scratch the surface. Digital TDM has a long way to go before it exhausts its potential. From the days when 8 Mbit/s transmission was regarded as state of the art, to the present when 565 Mbit/s multiplexers are considered routine, component technologists and circuit designers have managed to keep pace with the ever-increasing demands for telecommunications capacity. At present, multiplexing techniques are aiming towards the 10 Gbit/s target which seems to be the next hurdle for TDM. However, at these bit rates it is no longer a one-horse race, and competing technologies such as wavelength division multiplexing (WDM) on optical fibres are beginning to appear as economic alternatives. It is ironic that technology has also turned full circle, since

WDM is re-employing many of the techniques of FDM, this time on glass fibre rather than copper cable.

The other significant trend that will be observed is the convergence of transmission and switching products into one family interconnected by standard optical interfaces. This means that multiplexers will no longer be standalone components in a network, but rather they will become yet another logical function within a larger combined switching transmission element.

References

Ballart, R. and Ching, Y-C. (1989) 'SONET: Now it's the standard optical network', *IEEE Communications*, vol. 27, no. 3, pp. 8–15.

Brooks, R. M. *et al.* (1985) 'British Telecom 565 Mbit/s Lightline System', *British Telecom Tech. J.*, vol. 3, no. 2, pp. 46–51.

Brooks, R. M. *et al.* (1986) 'A highly integrated 565 Mbit/s optical fibre system', *British Telecom Tech. J.*, vol. 4, no. 4.

CCITT (1988a) 'G.736, Characteristics of synchronous digital multiplex equipment operating at 2048 kbit/s', in *Blue Book*, ITU, Geneva.

CCITT (1988b) 'Section 7.3, Characteristics of primary multiplex equipment', G.731–739, 'Section 7.4, Characteristics of 2nd order multiplex equipment', G.741–747, 'Section 7.5, Characteristics of higher order multiplex equipments', G.751–755, in *Blue Book*, ITU, Geneva.

CCITT (1990) 'G.707, Synchronous digital hierarchy bit rates', 'G.708, Network node interface for the synchronous digital hierarchy', 'G.709, Synchronous multiplexing structure', in *Blue Book*, ITU, Geneva.

Chu, W. W. (1969) 'Design considerations of statistical multiplexers', in *Proc. 1st ACM Symp. Prob. L Optimisation Data Comm. Syst.*

Chu, W. W. (1971) 'Design considerations of statistical multiplexers', in *Proc. 2nd ACM Symp. Prob. L Optimisation Data Comm. Syst.*

de Prycker, M. (1991) *Asynchronous Transfer Mode*, Ellis Horwood, London.

Hall, R. D. and McCartney, P. J. (1987) 'A 2 Mbit/s–140 Mbit/s synchronous multiplex system', *British Telecom Tech. J.*, vol. 5, no. 3, pp. 25–31.

Helmes, T. (1989) 'Crash-proof fiber rings', *Telephony*, vol. 217, no. 9.

IEEE (1990) 'Global deployment of SDH—compliant networks', *IEEE Comms.*, special edn, vol. 28, no. 8.

Newall, C. (1922) 'Synchronous transmission systems', Northern Telecom, Doc-GH9 Issue 3.

Smith, D. R. (1985) *Digital Transmission Systems*, pp. 127–88, Van Nostrand Reinhold, London.

Whitt, S., Hawker, I. and Callaghan, J. (1990) 'The role of Sonet-based networks in British Telecom', *IEEE Int. Conf. on Comms.*, 16 April.

7 Data synchronization

EDWIN JONES

In a digital transmission system, time synchronization usually has to be established in a number of respects, classified as follows:

- *Bit synchronization* refers to the data clock which must be re-established at a digital receiver in order to time the decision processes. This clock will usually be derived from the incoming pulse stream and so we use the expression *clock extraction.*
- *Word/frame synchronization* is concerned with locating the repetitive structure into which the data pulses are assembled prior to transmission. These data pulses will usually be formed into words, and if several information streams are to be combined into a time division multiplexed format before transmission, a frame structure will result. In addition, some modern communication systems data words are grouped together into cells or blocks before being inserted into the frame structure. There is usually a simple relationship between the word/cell/block structure and the frame structure such that once *frame alignment* has been established, then the other structural boundaries will also be known.
- *Carrier synchronization* is required if a data stream is to be modulated onto a carrier for transmission over a bandpass channel and a synchronous demodulation process is to be used at the receiver. As with clock extraction, the carrier has to be derived from the incoming modulated data stream.
- *Network synchronization* refers to issues concerned with the relative timing between terminals and nodes within a communication network. Questions arise as to whether all terminals should be synchronized to common timing signals and if so, how can these signals be reliably distributed about the network? Conversely, if communication links are allowed to run asynchronously, is it possible to ensure that data is never lost?

Clearly, synchronization is an important topic; if synchronism is lost, then data may also be lost. In this chapter we establish techniques by focusing on the first two categories, that is, on clock extraction and on frame alignment. Carrier synchronization for modulated systems is an integral part of the demodulation process and so is best dealt with as such. Likewise, network synchronization is more appropriately dealt with as a system design issue and so is addressed elsewhere in relation to multiplexing and networking topics.

7.1 Clock extraction

7.1.1 Self-timed systems

Clock extraction (known also as timing recovery or bit synchronization) is the process by which a digital regenerator obtains a synchronizing signal which enables it to optimize the timing of its decision-making process. (Decision-making and its relationship with the eye of the received signal has been discussed in previous chapters.)

Most digital transmission systems are *self-timed* in that they extract the clock from the incoming data stream. This avoids the need for a separate timing channel, as used in most computers for example, and makes it easier over the longer transmission distances involved to maintain the crucial phase relationship between the data and the clock at the point of decision-making.

Self-timing requires that the data signal is coded to ensure either that there is a clock component present in the transmitted signal or that such a component can be reliably extracted after processing at the receiver. The words 'reliably extracted' imply that a satisfactory clock can be recovered with a frequency, phase and amplitude adequately immune not only from transmission distortions and interference, but also from any signal-dependent effects arising from different data sequences. The latter requirement is found to be especially pertinent in transmission systems containing a number of regenerative repeaters in tandem. Each repeater experiences some pattern-dependent variation or jitter on the phase of its extracted clock and this is passed on via the regenerated data stream to the next repeater. Thus the locally generated phase jitter will combine with the incoming signal jitter, a situation which can lead to a progressive build-up of jitter at subsequent regenerators. Careful design is required to ensure that this potential timing problem remains within acceptable bounds.

7.1.2 Timing content of a digital signal

Fundamental to an explanation of clock extraction is a demonstration that all pulse sequences have symmetries in the amplitude and phase of their spectral structure. Even if the received signal lacks a component at the clock frequency, this spectral symmetry, which is independent of sequence statistics, can usually, after further processing, be used to provide a clock for retiming purposes.

Figure 7.1 illustrates, by way of examples, the spectral content of a number of continuously repeated seven-bit patterns of impulses. (The

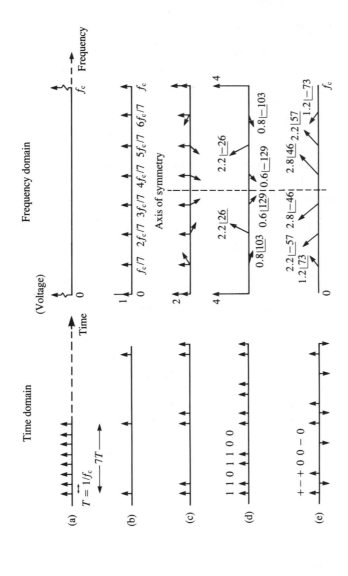

Figure 7.1 Repetitive sequences—time and frequency domain representation: (a) impulses at clock rate; (b) impulses at sequence rate (reference for amplitude and phase comparisons); (c) adjacent impulses; (d) arbitrary seven-bit word (1101100); (e) AMI for sequence 1110010

only significance in the choice of a seven-bit sequence is that, with seven
being prime, it does not give rise to additional structure which could
cloud the issues being illustrated.) Figures 7.1(a) and (b) define,
respectively, clock and sequence marker reference signals with which to
compare other sequences. If the clock frequency is f_c ($= 1/T$) then,
using Woodward's notation, we have a Fourier transform pair of:

$$\text{rep}_{nT}\,\delta(t) \leftrightarrow \sum_{i=-\infty}^{\infty} \delta\left(f - \frac{if_c}{n}\right) \qquad (7.1)$$

where i is the harmonic number,
$\quad n = 1$ for the clock signal of Fig. 7.1(a), and
$\quad n = 7$ for the sequence marker of Fig. 7.1(b).

If a second impulse is now introduced into the time sequence of Fig.
7.1(b) at time T after the first impulse, we get:

$$\text{rep}_{7T}(\delta(t) + \delta(t - T)) \leftrightarrow \sum_{i=-\infty}^{\infty} (1 + e^{-j2\pi f/f_c}).\delta\left(f - \frac{if_c}{7}\right) \qquad (7.2)$$

as shown in Fig. 7.1(c), where the phase of each additional spectral
component is plotted relative to the vertical component of the reference
signal of Fig. 7.1(b). (That is, the amplitude and phase of each spectral
component has been drawn as a vector with origin at the appropriate
point on the frequency axis.) It is seen that there are even amplitude and
odd phase symmetries, known as *Hermitian symmetry*, about $f_c/2$.

In fact, by generalizing Eq. (7.2) it follows that any repetitive impulse
sequence, including the important case where the sequence length n
tends to infinity (i.e., the random data stream), will yield this symmetry
condition. The condition is also replicated about k times $f_c/2$ for all
integer values of k. Figure 7.1(d) gives a further example with the
computed resultant phasors indicated on the frequency domain plot.
These spectral symmetries, which are independent of sequence statistics,
arise from the periodic structure associated with the signal clock. This
means that a random pulse sequence exhibits periodic or *cyclostationary*
statistics (Gardner and Franks, 1975). A more formal treatment of this
result is to be found in O'Reilly (1984).

All the sequences illustrated so far can be seen to contain a clock
component at frequency f_c. Thus, clock extraction could be achieved
simply by selecting this component with a bandpass filter. However, this
method is not satisfactory if the f_c component is weak or nonexistent.
The latter case will always arise if the sequence has no d.c. component;
this follows directly from the symmetry property noted above. Thus
most transmission codes, which deliberately aim to suppress any d.c.
component in the data sequence, will fall into this category. Figure
7.1(e) illustrates this observation for an alternate mark inversion (AMI)

coded sequence. (The AMI line code is described in Chapter 10.) For
such sequences, further processing is required before a clock component
can be extracted. This can be achieved by subjecting the received signal
to a nonlinearity such as rectification, clipping or squaring or, often in
practice, a combination of these. We thus note that transitions in a data
sequence (ones-to-zeros or vice versa) may be insufficient to guarantee a
clock component directly. The transitions are needed to ensure that
timing information is present, but nonlinear processing may be necessary
to extract the clock itself.

The appendix at the end of this chapter analyses the squarer, where it
is demonstrated that for any (ideal) impulse sequence a squaring process
will generate the requisite constant phase discrete spectral line at the
clock frequency f_c. For readers familiar with visualizing convolution in a
graphical manner, we can confirm this result by inspection of Fig. 7.1.
Squaring in the time domain corresponds to self-convolution in the
frequency domain. The self-convolution of the spectrum of Fig. 7.1(e)
can be seen to yield (at frequency f_c) pairs of spectral components with
equal amplitude but complementary phase. That is, they produce
resultants with identical phase (pointing vertically in the notation of Fig.
1) and so combine in a constructive manner to produce a nonzero
spectral line at frequency f_c.

7.1.3 Clock extraction circuits

A commonly used method of clock extraction for data signals which
have little or no spectral energy at the clock frequency is shown in Fig.
7.2. Section 7.1.2 discussed how a signal spectrum with Hermitian
symmetry about $f_c/2$ will, after nonlinear processing, produce a spectral
component at the clock frequency. Ideally, to ensure this symmetry
property, it is necessary to have a separate transmission channel
equalizer for the timing path. In practice, the equalizer of Fig. 7.2 is
often optimized for the decision path as discussed in Chapter 1 and
provides little more than noise bandlimiting for the timing path.
Nevertheless, a Hermitian symmetry component will be present at point
A, even if it is diluted by the characteristics of the transmitter and
channel.

The nonlinear network then produces, at point B, a discrete spectral
line at the clock frequency, f_c, together with a continuous spectrum
arising from the random data pattern. The purpose of the *clock selecting
filter* is to select the wanted clock while rejecting as much as possible of
the pattern-related spectrum together with any associated noise. This
selecting filter may be a narrow bandpass filter or a phase-locked loop
with effective Q factors ranging in practice from about 80 to in excess of
200. The choice of Q will depend critically upon the stability of the
incoming data rate and the tuning accuracy of the filter. A high Q

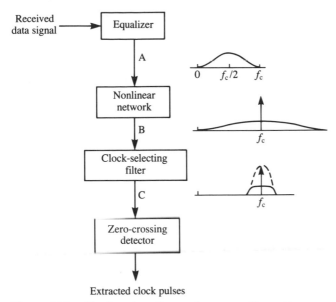

Figure 7.2   Clock extraction using a nonlinearity

minimizes the unwanted spectral components but gives rise to large phase shifts in the extracted clock if the tuned frequency deviates from the incoming clock frequency f_c.

Finally, the signal at point C is applied to a zero-crossing detector or hard limiter which removes the amplitude variations and gives clock pulses for use in the decision-making and regenerating circuits.

7.1.4 Jitter

Short-term variations from the optimum timing of a digital decision-making process are known as 'jitter'. It can be regarded as a phase modulation of the extracted clock relative to the original system clock and as such is sometimes referred to as *phase jitter*. The effect of jitter on the decision-making process within a regenerator is illustrated in Fig. 7.3(a), where it can be seen that clock edge jitter can cause decisions to be made at suboptimum times, thereby increasing the probability of error. This jitter is caused by a combination of:

- incoming noise affecting the extraction process, and
- data-dependent timing effects arising from imperfections and limitations in the clock extraction process.

Both of these effects arise from the unwanted continuous spectral components getting through the clock selecting filter to point C in Fig. 7.2. They combine with the spectral component of the wanted incoming clock to produce a resultant output clock which exhibits both amplitude

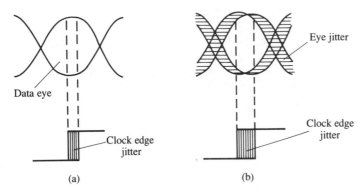

Figure 7.3 Timing jitter and the data eye: (a) clock edge jitter; (b) effect of jitter accumulation

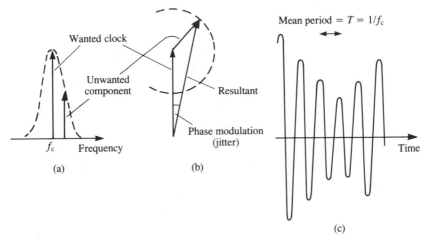

Figure 7.4 Jitter arising from an interfering spectral component: (a) spectrum from clock-selecting filter (Fig. 7.2, point c); (b) phasor representation of (a); (c) resulting waveform with amplitude and phase modulation

and phase modulation. The mechanism for a single interfering component is shown by way of example in Fig. 7.4. Part (a) shows the wanted and unwanted spectral components as they might appear at the output of a clock selecting filter. The equivalent phasor representation is given in part (b), where it is seen that the resultant has both amplitude and phase modulation. This is again apparent in the corresponding waveform of part (c). The amplitude variations can be effectively removed by a zero-crossing detector, provided that the clock component is always adequately large. However, the phase modulation will remain, giving rise to timing jitter.

In practice, appropriate noise bandlimiting at the regenerator input means that the contribution to jitter from the noise is usually small, i.e., the data-dependent timing effects tend to dominate. These effects can be removed if the spectrum at the output of the clock-selecting filter also has Hermitian symmetry, but now with a symmetry centred on f_0 (Franks and Bubrowski, 1974). This can be explained by considering the phasor diagram of Fig. 7.4(b). If the unwanted components always appeared in pairs with Hermitian symmetry centred on f_c, then each symmetrical pair would produce a vector of equal amplitude but opposite phase with respect to the unwanted clock. These would rotate about the f_c vector in opposite directions and so would produce a resultant that had amplitude modulation but no phase modulation, i.e., no jitter.

It has already been noted that any data sequence (in impulse form) has Hermitian symmetry properties which, after squaring, produce the requisite Hermitian symmetry about f_c. It thus remains to provide equalization to compensate for finite width and channel characteristics to ensure that the spectral symmetry condition is preserved at the output of the clock-selecting filter.

Unfortunately, this requirement is not consistent with the *skew symmetry* spectral requirements for a good received eye (Chapter 1) and so must be provided by separate spectral shaping within the clock path. Although this demonstrates a theoretical requirement, in practice a separate clock path equalization is not always found to be necessary. Acceptably low jitter performance can often be obtained by a judicious combination of the following: careful transmission coding to ensure an adequate clock component relative to the continuous spectrum after nonlinear processing (Jones and Zhu, 1987); received noise limiting; and an appropriate choice of Q for the clock-selecting filter.

7.1.5 Jitter accumulation

Jitter arising from the clock extraction and retiming process in a digital regenerator will be transmitted to subsequent regenerators, where it may combine with any locally generated timing variations to cause an overall accumulation of jitter. When assessing the effect of this jitter accumulation in a multiregenerator self-timed system, two interrelated factors have to be taken into account. These are *eye jitter* and *clock edge*

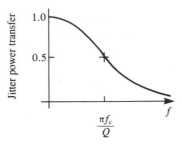

Figure 7.5   Jitter transfer characteristic for a tuned circuit

jitter and are depicted in Fig. 7.3(b). Eye jitter is a direct consequence of any timing variations already present on the incoming data signal. Clock edge jitter arises as described in Sec. 7.1.4 (which refers to Fig. 7.3a) except that now a contribution will also be made to it by the incoming data signal.

The decision-making process at the regenerator may not be as difficult as implied by Fig. 7.3(b). The extracted clock will tend to follow any phase variations in the incoming data. What matters as far as the instantaneous signal-to-noise ratio and so decision error probability is concerned, is the *relative* timing of a particular decision instant. That is, the difference between the time position at the clock edge and the optimum decision time for the particular incoming data symbol. Although this jitter tracking may help the decision-making process at a particular regenerator, jitter will still accumulate from one generator to another. Problems can arise at the end of a link where the data signal may have to relate to a more rigid *network clock*. At this point, the incoming data signal may contain timing jitter which amounts to several clock periods. If data slips (loss or gain of bits) are to be avoided, quite large data buffers may be required. The way in which jitter builds up has been extensively studied in the literature (see, for example, Byrne *et al.*, 1963). Suffice it to say here that it is the data pattern-dependent effects which cause most trouble, as they tend to reinforce each other at subsequent repeaters. In long chains of regenerators, it is usually preferable to prevent the build-up of excessive jitter in the first place by using carefully designed phase-locked loops rather than band-pass filters in the clock selection circuits. Alternatively, data scramblers may be installed at intervals to break up the data patterns and so control the main contributor to the problem, namely, the build-up of data-dependent jitter.

The amount of jitter passed through a regenerator will depend upon the detail of the clock extraction process and, in particular, on the parameters of the clock-selecting filter. For a simple tuned circuit the jitter power transfer characteristic (Byrne *et al.*, 1963) will depend upon the Q of the filter as shown in Fig. 7.5. Thus, low-frequency jitter will be

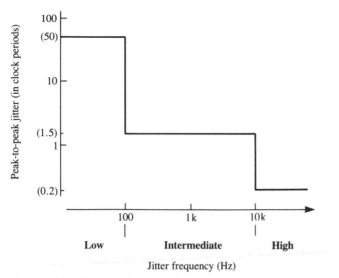

Figure 7.6   Typical jitter limits for a digital transmission system

transferred through a regenerator with very little attenuation, while higher frequency jitter will be reduced. It follows that any jitter suppressors or data buffers need to be designed to deal especially with low-frequency jitter, thus they may need to store large numbers of consecutive data bits so that any timing jitter can be smoothed out.

7.1.6 Jitter specifications

Specifications for jitter accumulation in digital systems have to take account of the low-frequency build-up implied in Figure 7.5. A typical specification for jitter limits for a digital transmission system comprising many regenerators is given in Fig. 7.6 where peak-to-peak jitter, expressed in clock periods, is plotted against jitter frequency. (It should be noted that r.m.s. values are typically a factor of 10 less than these figures.) This representative example is loosely based on figures given in CCITT recommendations G.823 and G.824, although reference to these will show that the higher levels in the multiplexed transmission hierarchy attract somewhat tighter specifications. Figure 7.6 serves to highlight the low-frequency timing jitter (or wander) which can amount to many clock intervals.

7.1.7 Further reading

Apart from the references given in the text, useful chapters on clock extraction are to be found in Bylanski and Ingram (1986) and also in Lee and Messerschmitt (1988); the latter includes details of waveform sampling methods which are applicable to fully digital realizations. For detailed and analytical treatment, the early work on timing and jitter published in the *Bell System Technical Journal* should be consulted (in particular the papers by Bennett and by Rowe, both in vol. 37, 1958, and also the paper by Manley in vol. 48, 1969). On jitter in digital networks, Kearsey and McLintock (1984) is a helpful paper which makes particular reference to the CCITT recommendations.

7.2 Frame alignment

7.2.1 The task

In addition to clock extraction, most signal formats used for digital transmission will entail some further structure which must be reliably extracted at the receiving terminal. In particular, when several information streams are combined before transmission using time division multiplexing, a frame structure will be involved. The boundaries between frames are usually marked by inserting a carefully selected frame alignment word (also called a marker, flag or frame sync pattern) which the receiver has to locate before demultiplexing can be performed (Fig. 7.7).

The reliable frame alignment of such multiplexed signals is essential for the proper functioning of many digital communication systems. A loss of alignment results in the failure to identify correctly the received bits and so causes the disorientation of the demultiplexing process. This leads to a catastrophic loss of both message and control information. Thus the choice of frame alignment word (FAW) and the design of reliable *alignment detection*, *misalignment detection*, and *searching* algorithms is crucial. Figure 7.8 shows the principal tasks which have to be performed, with the desirable state shown in bold.

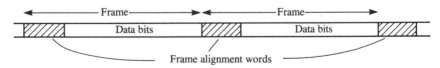

Figure 7.7   Frame structure

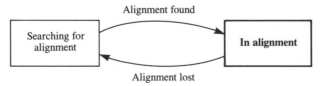

Figure 7.8   The alignment task

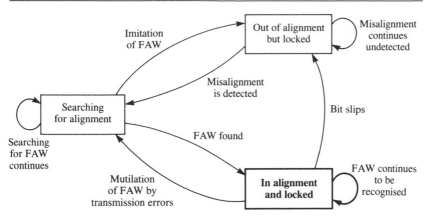

Figure 7.9 A state transition diagram for frame alignment

In practice, the aligner's task is made yet more challenging when one notes that:

1 The frame alignment word may, in most systems, be temporarily imitated by a sequence of data bits in an arbitrary position within the data region.

2 The frame alignment word may be mutilated by transmission errors.

3 In some systems bit slips can occur, that is, incoming frames may temporarily lose bits or acquire erroneous bits.

To deal with these problems the two-state model of Fig. 7.8 has to be extended to at least the three-state model of Fig. 7.9. This model distinguishes between two locked (not searching) states: the desirable 'in alignment and locked' state and the highly undesirable 'out of alignment but locked' state. In this latter state a misalignment condition has occurred but the system has not yet detected it. The causes for transition from one state to another are given in the diagram; it can be seen that they relate directly to the conditions 1–3 listed above.

Before considering the mechanisms involved in such an alignment process, we note that there are variations in the above which can be found in practice. The FAW may not be transmitted in the *bunched format* shown in Fig. 7.7. If a system is known to be prone to short bursts of transmission errors, some protection against them may be afforded by distributing the FAW in known positions throughout the frame. Whichever method is adopted, the alignment principles are similar. Secondly, some systems use a unique FAW, i.e., the data bits are specially encoded so that in the absence of transmission errors the FAW can never be imitated (e.g., the frame marker flag 01111110 in CCITT recommendation X.25). For a given level of confidence in the alignment process, this simplifies the requisite algorithms but at the expense of some redundancy in coding the data bits.

7.2.2 The alignment process

When a system is operating correctly the aligner should reside in the 'in alignment and locked' state of Fig. 7.9, with the equipment verifying that the FAW occurs in the predicted position in each frame. For most well-behaved transmission systems, this is the state that is occupied for a high proportion of the time. If a misalignment condition is then detected, the mechanism must transfer to the searching state and begin an inspection of all possible FAW positions. When the correct position has been found, the search can be abandoned and the equipment allowed to revert as quickly as possible to the locked condition.

In practice, the alignment mechanism needs to be yet more complicated than that shown in Fig. 7.9. This is because it is desirable to try and verify that the current state really is incorrect before transferring to another state. For example, when the equipment is 'in alignment and locked', transmission errors may occasionally mutilate the FAW in a given frame; it may be sensible to wait and check the next frame (or the next few frames) to minimize the probability of setting off on an unnecessary search. Similarly, when in the 'searching' state, a match to the FAW may be found which, in fact, is an imitation caused by the data bits. This condition is unlikely to be duplicated in the same position in successive frames (unless the FAW chosen is a bad one!) and so again, a check on subsequent frames may be sensible before reverting to the locked condition. Thus, in practice, a series of *check states* are inserted between the locked and searching states of Fig. 7.9 to reduce the probability of erroneous transitions between states. Of course a compromise is necessary if a genuine transition to another state is required, it is equally undesirable to waste time making unnecessary checks. Unfortunately, the equipment cannot instantaneously distinguish between necessary and erroneous transitions. Table 7.1 gives some examples of practical frame alignment arrangements; the last two columns detail the recommended number of adjacent frames that should be inspected before changing state.

To summarize, when in alignment, a good frame alignment process should exhibit a low probability of losing alignment and, when out of alignment, a high probability of fast recovery to the aligned condition. In order to achieve these objectives the following features are required:

- Reliable verification of the in-alignment condition
- An efficient search procedure which ensures rapid location and verification of the position of the FAW
- Rapid detection of a genuine out-of-alignment condition caused by bit slips
- Robust performance in the presence of FAW imitations and transmission errors

Table 7.1 CCITT recommendations for frame alignment (for the 30 telephone channel-based multiplex hierarchy)

Multiplex level	CCITT recommendation	Frame length (bits)	Frame alignment word	Number of frame checks before changing state	
				Lock-to-search	Search-to-lock
Primary ($\approx$ 2 Mbit/s)	G.732	512	0011011 (7 bits)	3 or 4	3
Second ($\approx$ 8 Mbit/s)	G.745	1056	11100110 (8 bits)	5	2
Third ($\approx$ 34 Mbit/s)	G.751	1536	1111010000 (10 bits)	4	3
Fourth ($\approx$ 140 Mbit/s)	G.751	2928	111110100000 (12 bits)	4	3

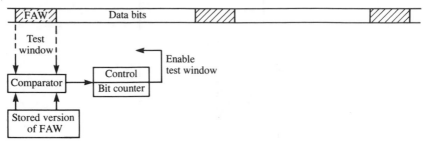

Figure 7.10   Basic configuration for a frame aligner

Ideally these features should be obtained with a minimum of added redundancy in the data stream (in the form of an FAW) and with reasonably simple alignment equipment.

Inevitably, there will be some compromises here. An FAW that is short compared with its frame length will make for a low redundancy system but will increase the opportunities for misinterpretation of the frame position (through imitation of the FAW in the data region). Waiting to check FAWs in successive frames and/or allowing a limited number of transmission errors in an 'acceptable' FAW may yield a better system, but of course at the expense of added complexity.

7.2.3 Searching techniques

A simple bit-by-bit aligner is outlined in Fig. 7.10 where a *test window*, of length equal to the FAW, is shown to be correctly aligned with the incoming signal. That is, a *comparator* (or pattern matcher) indicates that there is agreement between the received FAW and a locally stored version. This causes a bit counter to count through the known frame length and so locate and verify the position of the next FAW, and so on. If an FAW is not found in the expected position the control circuit will initiate a bit-by-bit search through the frame until a new alignment position is found.

As discussed in Sec. 7.2.2, it will usually also be necessary to guard against erroneous action being taken when a particular received FAW is mutilated by transmission errors or when an imitation of an FAW occurs within the data region. We have seen that this can be done by checking an appropriate number of adjacent frames before entering or leaving the search state. A further sophistication is that of *off-line searching* in which extra equipment commences a search as soon as a misalignment condition is suspected. This takes place in parallel with the normal receiver processing and offers the considerable advantage that if the search proves to have been unnecessary no damage will have been done. It also benefits the true out-of-alignment condition in that an early start will have been made on the searching process.

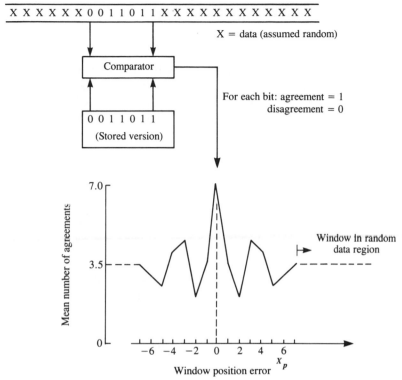

Figure 7.11   Mean number of bit agreements for different window positions

Simple bit-by-bit searching techniques can be relatively slow in establishing the aligned condition when there are random FAW imitations within the data region. This arises because each time an imitation is found, the search stops and a check is made for an FAW one frame later. On failing to find the requisite pattern, the search resumes. This temporary halt in the searching process occurs every time an FAW imitation is found. A faster search can be achieved by using more memory. For example, the location of sequences which match the FAW can be recorded and the search allowed to continue. Those which do not repeat at one-frame intervals are ignored. This process continues until the location of the true FAW becomes apparent. The aligner can then be made to jump directly to this new position.

Thus, there is a trade-off; faster searches are possible but at the expense of equipment complexity.

7.2.4 Choice of frame alignment word

Some recommended patterns have been given in Table 7.1. By way of example, Fig. 7.11 takes the first pattern in the table (for a 30-channel p.c.m. system) and plots the mean number of bit matches from the alignment comparator for different window positions. When the test window is in the correct position (and there are no transmission errors) the comparator registers seven agreements. If the position of the window is in error by one bit (in either direction), then it can be seen that three agreements and three disagreements are registered, with the seventh bit overlapping into the data region. For random data, this last bit registers an agreement or disagreement with equal probability, thus overall, the mean number of agreements for a window position error of one bit is 3.5. Repeating this calculation for other window positions generates the plot of Fig. 7.11. This confirms that the sequence 0011011 is a good FAW in that it produces a strong correlation spike when in alignment with relatively low sidelobes associated with the out-of-alignment window positions. In this respect we note that the correlation values for alignment errors of one or two bits are especially low, i.e., it is a good word to use for well-behaved systems which rarely experience more than one or two bit slips. Under such circumstances the pattern is thus seen to be reasonably immune to unfavourable data bit combinations and/or isolated transmission errors, both of which could increase the apparent number of agreements but are unlikely to achieve the in-alignment peak value.

The choice of FAW for a particular system depends upon many factors. For example, in contrast to the above well-behaved application, systems which have to acquire alignment frequently, particularly from a random starting position, usually require longer FAWs. On the other hand, long FAWs are more liable to experience transmission errors. Thus, the word choice and the alignment strategy are closely linked. For example, in the presence of transmission errors it may be worth while to accept window positions that demonstrate a close but not necessarily exact match to the expected FAW. The implications of this and other variations on the basic alignment approach described above may be found in the literature.

7.2.5 Further reading

A good introduction to frame alignment is to be found in Bylanski and Ingram (1986) and in several more recent books which refer to them and adopt a similar approach. Some of the subtleties of the alignment process with examples based on CCITT specifications are discussed in Jones and Al-Subbagh (1985); a similar method of analysis (using probability generating functions) is adopted in a tutorial article by Choi

(1990). Barker (1953) was first to study the selection of good frame alignment words, since then many variations have been investigated to provide for optimum performance under a variety of system conditions; a more recent study with references to earlier work is to be found in Al-Subbagh and Jones (1988).

Appendix   Generation of a clock component using a square law device

From the time-frequency transform pairs of Fig. 7.1, it follows that any impulse sequence can be seen as a sum of cosine terms (harmonics) of amplitude a_i and phase ϕ_i, thus:

$$f(t) = \sum_{i=-\infty}^{\infty} a_i \cos\left(2\pi \frac{if_c}{n} t + \phi_i\right) \qquad (A.1)$$

On squaring, we can write:

$$f^2(t) = \sum_{i=-\infty}^{\infty} \sum_{j=-\infty}^{\infty} a_i \cos\left(2\pi \frac{if_c}{n} t + \phi_i\right) \cdot a_j \cos\left(2\pi \frac{jf_c}{n} t + \phi_j\right) \quad (A.2)$$

Using the trigonometrical relationship $\cos A \cos B = \frac{1}{2}[\cos(A + B) + \cos(A - B)]$:

$$f^2(t) = \sum_{i=-\infty}^{\infty} \sum_{j=-\infty}^{\infty} \frac{a_i a_j}{2}\left[\cos\left(2\pi \frac{(i+j)f_c}{n} t + \phi_i + \phi_j\right)\right.$$

$$\left. + \cos\left(2\pi \frac{(i-j)f_c}{n} t + \phi_i - \phi_j\right)\right] \quad (A.3)$$

Terms contributing to a clock component at frequency f_c will arise when $i + j = n$ in the first part of Eq. (A.3) and when $i - j = n$ in the second part, i.e.:

$$f^2(t)\Big|_{f_c} = \sum_{i=-\infty}^{\infty}\left[\frac{a_i a_{n-i}}{2} \cos(2\pi f_c t + \phi_i + \phi_{n-i})\right.$$

$$\left. + \frac{a_i a_{-(n-i)}}{2} \cos(2\pi f_c t + \phi_i - \phi_{-(n-i)})\right] \quad (A.4)$$

Now for a cosine function $a_{n-i} = a_{-(n-i)}$ and $\phi_{n-i} = -\phi_{-(n-i)}$, therefore:

$$f^2(t)\Big|_{f_c} = \sum_{i=-\infty}^{\infty} a_i a_{n-i} \cos(2\pi f_c t + \phi_i + \phi_{n-i}) \qquad (A.5)$$

From our generalization of the examples of Fig. 7.1 we have noted that any impulse sequence has Hermitian symmetry about $f_c/2$, i.e.: $a_i = a_{n-i}$

and $\phi_i = -\phi_{n-i}$, thus:

$$f^2(t)\Big|_{f_c} = \sum_{i=-\infty}^{\infty} a_i^2 \cos 2\pi f_c t \tag{A.6}$$

that is, the squaring process has generated contributions to a clock component, all of which, because of the Hermitian symmetry property, have the same phase. A more rigorous proof of this result is to be found in Blachman and Mousavinezhad (1990).

In the above discussion each symbol has been assumed to be an impulse. In practice, symbols have finite width and may also be subject to dispersion (in a channel for example). If the resulting effect is to produce a symbol $s(t)$, then it is necessary to convolve this function with the time domain waveforms of Fig. 7.1. For example, the generalized form of Eq. (7.2) becomes:

$$s(t)*\text{rep}_{nT}(\delta(t) + \delta(t - T)) \leftrightarrow S(f) . \sum_{i=-\infty}^{\infty} (1 + e^{-j2\pi f/f_c}) . \delta\left(f - \frac{if_c}{n}\right)$$

$$\tag{A.7}$$

that is, the spectrum is multiplied by $S(f)$. This will not eliminate the symmetry properties but it will usually dilute them. Timing path equalization can be used to alleviate this effect, although in practice it is seldom considered necessary.

References

Al-Subbagh, M. N. and Jones, E. V. (1988) 'Optimum patterns for frame alignment', *Proc. IEE (Part F)*, vol. 135, 594–603.

Barker, R. H. (1953) 'Group synchronising of binary digital systems', in *Communication Theory*, W. Jackson (ed.), 273–87, Academic Press, New York.

Bennett, W. R. (1958) 'Statistics of regenerative digitial transmission', *Bell Syst. Tech. J.*, vol. 37, no. 6, 1501–1542.

Blachman, N. M. and Mousavinezhad, S. H. (1990) 'The spectrum of the square of a synchronous random pulse train', *IEEE Transactions on Communications*, vol. 38, 13–17.

Bylanski, P. and Ingram, D. G. W. (1986) *Digital Transmission Systems*, 2nd edn, Peter Peregrinus, London.

Byrne, C. J., Karafin, B. J. and Robinson, D. R. (1963) 'Systematic jitter in a chain of digital regenerators', *Bell System Technical Journal*, vol. 42, 2679–714.

Choi, D. W. (1990) 'Frame alignment in a digital carrier system—a tutorial', *IEEE Communications*, vol. 28, no. 2, 47–54.

Franks, L. E. and Bubrowski, J. P. (1974) 'Statistical properties of timing jitter in a PAM timing recovery system', *IEEE Transactions on Communications*, vol. 22, 913–20.

Gardner, W. A. and Franks, L. E. (1975) 'Characterisation of cyclostationary random signal processes', *IEEE Transactions on Information Theory*, vol. 21, 4–14.

Jones, E. V. and Al-Subbagh, M. N. (1985) 'Algorithms for frame alignment—some comparisons', in *Proc. IEE (Part F)*, vol. 132, 529–36.

Jones, E. V. and Zhu, S. (1987) 'Data sequence coding for low jitter timing recovery', *Electronics Letters*, vol. 23, 337–8.

Kearsey, B. N and McLintock, R. W. (1984) 'Jitter in digital telecommunications networks', *British Telecommunications Engineering*, vol. 3, 108–16.

Lee, E. A. and Messerschmitt, D. G. (1988) *Digital Communication*, Kluwer Academic Publishers, Boston.

Manley, M. M. (1969) 'The generation and accumulation of timing noise in PCM systems—an experimental and theoretical study', *Bell Syst. Tech. J.*, vol. 48, no. 3, 541–613.

O'Reilly, J. J. (1984) 'Timing extraction for baseband digital transmission', in *Problems of Randomness in Communication Engineering*, K W. Cattermole and J. J. O'Reilly (eds), Pentech Press, London.

Rowe, H. E. (1958) 'Timing in a long chain of regenerative binary repeaters', *Bell Syst. Tech. J.*, vol. 37, no. 6, 1543–98.

8 Errors and error extension

JEFFREY DESLANDES

The decision error mechanisms found in digital transmission were briefly introduced in Chapter 1, and discussed in more detail in Chapter 5. In this chapter we consider the consequences of such errors in digital transmission systems, including some causes of error activity and how it can be measured. The effect that errors can have on a transmission network itself is also discussed for both the established plesiochronous digital networks and for the new synchronous digital networks. We then consider how a circuit is routed through the transmission network, with the implications on performance monitoring and the generation of alarms as errors propagate.

8.1 The need for performance measures

Transmission errors are usually the most significant impairment to be found in a digital communication system. They are also difficult to predict and therefore difficult to quantify. Consequently, the study of their occurrence and the means of specifying their characteristics have been and continue to be the subject of considerable activity (see Yamamoto and Wright (1989) and references therein).

The principal causes of transmission errors in a communication system are:

- Noise (in particular, thermal noise for metallic cable and radio systems, and quantum noise for optical systems)
- Impulsive interference (often from nearby electromechanical equipment, lightning disturbances, or maintenance activities)
- Crosstalk and intermodulation (from other interfering transmission equipment)
- Echoes and signal fading (arising from mismatches in transmission channels and from multipath effects in radio systems)
- Terminal equipment limitations and maladjustments (such as errors in equalization and decision timing)
- Network problems (including frame synchronization errors and lost bits)

These effects can lead to error bursts ranging from tens of microseconds through to seconds. Furthermore, the disturbance experienced by a user will be dependent upon the type of service being carried by the circuit. For example, random errors and very short bursts can cause small disturbances to PCM signals which, in general, will not be detectable by telephony, although such errors may cause a significant number of retransmissions for data services. Whereas, if a comparable number of errors are grouped into long bursts, the number of data retransmissions will reduce, but there may be noticeable disturbances to speech. In addition, consideration must be given to the influence of error activity on the transmission network itself. This can result in corruption of the demultiplexer synchronization and justification mechanisms, and will be discussed later in this chapter.

Various methods have been used to model the causes of degradation and interference in digital transmission; a useful review is to be found in Kanal and Sastry (1978) while Knowles and Drukarev (1988) provides a more recent example. It is worth noting that at the receiving terminal, or at a regenerator in a digital link the probability of making an erroneous decision (the error probability) is critically dependent upon the received signal-to-interference ratio. A small change in this ratio will result in a large change in the error probability, so making digital systems 'brittle' in nature. It is therefore vital to establish error performance measures so that both users and transmission system providers can anticipate an end-to-end performance which is compatible with the requirements of the digital information to be carried. Such performance standards are now emerging, from the CCITT for example.

8.2 Measurement methods

Various methods have been used for measuring error activity within a digital system, each with its own limitations. This section introduces a number of methods and explains their uses and relative advantages.

8.2.1 Bit–error ratio

The *bit–error ratio* (BER) (CCITT, 1988b), sometimes referred to as the bit–error rate, is defined as: 'the number of bits received in error divided by the total number of bits transmitted in a specified time interval'. Within the specified interval, it is numerically equal to the bit–error probability.

Studies have been conducted to quantify the BER requirements for various digital services, see Yamamoto and Wright (1989), for example. Some typical objectives, expressed in terms of the *long-term mean BER*, are given in Table 8.1. However, the long-term mean BER is usually an

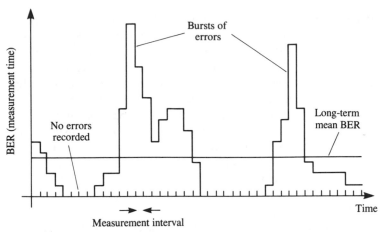

Figure 8.1   Variation of BER with time in a burst error environment

insufficient performance measure by itself since it provides no information on the arrival characteristics of the errors. As already discussed, the degree of 'burstiness' of the errors, that is, their distribution, is important for many digital services. Figure 8.1 shows a typical result where, for a given long-term mean BER, the BER over shorter measurement intervals is seen to exhibit considerable variation with time. During some measurement intervals no errors are recorded, while at other times bursts of interference can cause bursts of errors and thus a high *short-term BER*.

Table 8.1   Some typical BER objectives

Digital service	Transmission rate (approximate)	Long-term mean BER
Voice: log-law PCM	64 kbit/s	2×10^{-5}
ADPCM	32 kbit/s	10^{-4}
Video (broadcast standard):		
linear PCM	60 Mbit/s	2×10^{-7}
interframe coding	2 Mbit/s	$\sim 10^{-10}$
Data	16 kbit/s to 600 Mbit/s	$\sim 10^{-7}$

Within the illustrative limits quoted in Table 8.1, speech, video, and data services have quite different long- and short-term BER requirements. For speech, a listener is usually tolerant to the low level of background noise caused by randomly distributed errors (Takahashi, 1988). Burst errors usually cause more objectionable audible clicks but are still tolerable if they do not occur too often. Errors in a digital video signal can impair picture quality (CCIR, 1986) and again a viewer is usually happier with the effects of random rather than burst errors.

With video, however, additional care is needed as excessive bursts of errors can result in a sudden and catastrophic loss of picture synchronization. A similar sharp bound on acceptability may also occur with data if *error control coding* is used to protect the data from transmission errors. Occasional errors in an encoded block, or even bursts of limited duration, may be acceptable (in a retransmission scheme for example). However, a point will come when the error control coding system can no longer cope and then significant numbers of decoded data blocks may contain errors. At this point the proportion of *error-free blocks* becomes a useful measure of performance.

Thus, in order to quantify the error distribution characteristics, the BER measure must be refined by the addition of other measures.

8.2.2 Errored seconds

One method of improving the BER measure involves reducing the integration period to a small manageable unit, such as one second. This small integration period defines the resolution of the measure and provides information on the burstiness of the error activity. For example, the *error-free second* (EFS) parameter can be employed. This is defined as 'the proportion of one-second intervals, in a specified time interval, in which the transmitted data is delivered error free'. It is usually expressed as a percentage.

This approach can be further developed to provide information on the severity of the error burst by categorizing the value of the BER within each second. A typical breakdown may involve a measure for just a few errors and another for an excessive number of errors within each one-second period. These parameters are referred to as the *errored second* (ES) and *severely errored second* (SES) in the performance standard example quoted in Sec. 8.3.

For modern high bit rate transmission systems a second may be too long a period (as it will contain many bits) and so the *error-free millisecond, error-free microsecond,* or more specifically, an error free data block (as mentioned above) may become more appropriate measures.

8.2.3 Available time

If the error activity continues at an excessive level for a significant period of time (10 seconds in the example in Sec. 8.3), then the channel is considered to be unavailable. The measure used to monitor this condition is that of *available time* (or *availability*) and is defined as 'the percentage of time, in a specified time interval, in which the BER is less than a specified threshold'. This is typically quoted in excess of 99.8 per cent for a 2 Mbit/s circuit.

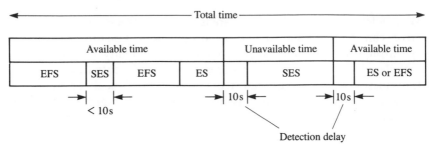

Figure 8.2 Relationship between error performance parameters

This measure is especially useful when blocks of data are protected by error control coding as discussed in Chapter 9. The coding algorithm can usually handle a given number of errors in a block. Blocks with more errors are decoded erroneously. If the BER measurement interval corresponds to the error control block length then, with the appropriate BER threshold, the available time will correspond to the percentage of error-free blocks after decoding the received data.

8.3 Performance standards—an example

The parameters BER, EFS, available time, and adaptations of them have been used in the emerging error performance standards for digital links (McLintock and Kearsey, 1984). The important CCITT recommendation G.821 (CCITT, 1988b) serves as an example. It specifies error performance objectives for long-distance 64 kbit/s data connections (Table 8.2).

Table 8.2 Error performance objectives—an example (CCITT, 1988a)

Performance parameter	Objective
Errored seconds (ES)	Fewer than 8% of one-second intervals to have any errors (i.e., EFS > 92%)
Severely errored seconds (SES)	Fewer than 0.2% of one-second intervals to have a BER worse than 10^{-3}

These are long-term objectives, typically measured over a period of one month, and they apply only to the available time of a connection. Figure 8.2 illustrates the relationship between the various performance parameters and a possible sequence of events leading to a period of unavailable time. The specification requires 10 consecutive SESs before the circuit enters the unavailable state. The unavailable time is then measured from the first of the 10 consecutive SESs. When the error activity reduces to produce 10 consecutive non-SESs (i.e., consisting of

ESs or EFSs) the circuit returns to the available state, which is again timed from the beginning of the 10-second period. Some subtleties of the terminology are to be found in the full specification (CCITT, G.821).

From the network operator's point of view, this specification is somewhat incomplete. With the increasing deployment of digital networks a *connection* often comprises a number of *links*, perhaps of very different type. The network operator needs to consider the effect of these individual links on the overall G.821 objectives which the customer expects to receive. Outline recommendation M.550 (CCITT, 1988c) attempts to address these matters by defining performance objectives for:

- The long term (several days)—for bringing a digital link into service
- The short term (several minutes/hours)—for removing an unsatisfactory link for maintenance

However, this is a complex and challenging topic which remains the subject of continuing study (Yamamoto and Wright, 1989; Kubat and Bollen, 1989).

8.4 Error extension mechanisms

In this section we consider the way errors propagate through a digital transmission network and how the underlying hierarchy can modify the characteristics of the original error activity. Other forms of error extension mechanism, such as from source and channel coding schemes, should be less significant if the system has been well designed and so are not considered here.

8.4.1 Plesiochronous digital hierarchy

By way of an example, let us consider the effect errors can have on the European plesiochronous digital hierarchy (PDH), although the results presented will, in general, apply equally to other versions of the PDH. We start with a brief description of the multiplexer frame structure as a reminder of the techniques employed. For a more complete overview of digital multiplexing schemes, see Chapter 6.

The European PDH consists of bit-interleaved multiplexers spanning 2–8–34–140 Mbit/s (CCITT, 1976; 1972b), with 64 kbit/s channels multiplexed to 2 Mbit/s using a synchronous word-interleaved process (CCITT, 1972a), see Fig. 8.3. The start of each frame is identified by the frame alignment word (FAW) to enable the demultiplexer to synchronize to the incoming frame and locate data belonging to each tributary. (The bulk of transmission across the UK network is currently performed at either 140 or 565 Mbit/s; note that the 140/565 Mbit/s multiplexer is part

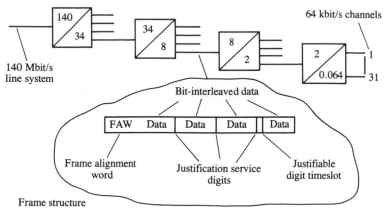

Figure 8.3   European plesiochronous digital hierarchy

of the 565 Mbit/s line system and will have a proprietary frame structure not considered within this chapter.)

As noted in Chapter 7 (Table 7.1) the CCITT frame alignment strategy for plesiochronous demultiplexers usually requires four consecutive FAWs to be corrupted before a loss of alignment condition is assumed and the realignment process commences. The demultiplexer must then find three consecutive error-free FAWs before realignment is achieved. When a demultiplexer has lost alignment it imposes the *alarm indication signal* (AIS), consisting of unstructured binary 1s, onto the tributary data until the realignment process has been completed.

The plesiochronous (or nearly synchronous) problem (see Chapter 6) is solved by allocating an extra timeslot to each tributary within the multiplexed frame. This is called the *justifiable digit timeslot* (or *stuffable digit timeslot*) and contains either valid tributary data, or is justified using a stuffed or dummy bit. The decision to justify is dependent upon the relative speeds of the tributary and multiplexer clocks, which operate at nominal frequencies within a tolerance, and may be subjected to jitter and wander. The demultiplexer determines the content of the justifiable digit timeslot by decoding the *justification service digits* (sometimes referred to as the *stuffing service digits* or *justification control digits*). This involves a majority decision process on three or five bits, depending on the multiplexer level. The justification service digits are distributed throughout the frame to improve their immunity to error bursts. If the majority of justification service digits become corrupted, the contents of the justification digit timeslot will be interpreted incorrectly, resulting in an *uncontrolled bit slip* passed on to the tributary. This will cause all lower order demultiplexers to lose alignment, unless a cancelling slip occurs first. For interested readers a useful description of the justification process can be found in Owen (1982).

As an example, let us further consider the 140–34 Mbit/s demultiplexer

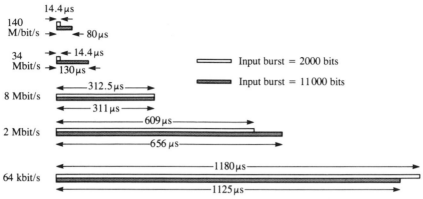

Figure 8.4   Example of error propagation through the PDH

which uses the above frame alignment strategy and employs a majority
decision on five justification service digits. The frame length of 21 μs
(2928 bits at 139.264 Mbit/s) requires a burst in excess of 63 μs to
corrupt four FAWs and cause the demultiplexer to lose alignment.
However, the justification service digits are separated by 488 bits and
only require a burst to be longer than 7 μs (976 bits) before the majority
decision can become corrupted, resulting in an uncontrolled bit slip.

A study of the error propagation mechanisms (Deslandes, 1991) has
confirmed that the effect of an error burst is dependent on both the
length and the bit-error probability of the burst. Short bursts with a
high bit-error probability (say 50 per cent) are likely to cause slips
directly, whereas lower bit-error probability bursts of longer length may
hit four consecutive FAWs, causing the demultiplexer to lose alignment
without corrupting the justification process. However, the cascaded
effect of the hierarchy will lead to a demultiplexer imposing an AIS on
its tributaries, which will result in a high probability of uncontrolled bit
slips being produced at lower levels of the hierarchy.

Software modelling the PDH—an example

A suite of programs can model the error performance of a demultiplexer
hierarchy, and the results obtained can then be verified both by
measurements on commercially available multiplexers (Deslandes *et al.*,
1990a) and by using theoretical analysis (Deslandes *et al.*, 1990b). Figure
8.4 shows an example where this technique has been applied to the
European PDH demultiplexers spanning 140–34–8–2–0.064 Mbit/s, with
two error bursts of differing types applied at 140 Mbit/s. The 2000 bit
burst (14.4 μs) is too short to cause the 140–34 Mbit/s demultiplexer to
lose alignment and results in the same length of error disturbance at
34 Mbit/s. However, it has produced an uncontrolled bit slip at
34 Mbit/s, causing loss of the 34–8 Mbit/s demultiplexer, which in turn

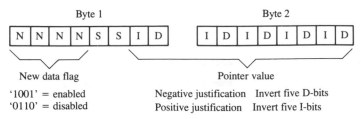

Figure 8.5 SDH pointer layout

causes loss of alignment of the hierarchy below. The burst of 11 000 bits (80 μs) causes loss of the 140–34 Mbit/s demultiplexer and has the same effect on the rest of the hierarchy. The likelihood of such error extension is, of course, dependent on the bit-error probability of the burst. However, this example illustrates the sensitivity of the PDH to relatively small bursts disturbing the justification process.

As an interesting exercise, the models can be modified to disable the justification process, resulting in a synchronous operation of the plesiochronous hierarchy. This is explained by Deslandes *et al.* (1990b) and has clearly identified the uncontrolled bit slip as the critical error propagation mechanism, in terms of both sensitivity to the input burst length and propagation of demultiplexer loss of alignment through the multiplexing hierarchy.

Although this example has been based on the European PDH, the error performance of other PDH hierarchies in operation outside Europe can equally be modelled using this technique. However, it would be expected that a similar set of results would be obtained.

8.4.2 Synchronous digital hierarchy

The synchronous digital hierarchy (SDH) has been developed by the CCITT (CCITT, 1990a,b,c) from the original ideas proposed by Bellcore for the synchronous optical network (SONET), see Boehm *et al.* (1985). The principles and notation of the SDH are described in Chapter 6. In this chapter we review the mechanisms for error extension to identify the SDH error performance when compared to the PDH.

The problems associated with synchronization of a large (or in fact global) network has resulted in the SDH not being totally synchronous. To accommodate any small frequency differences that may exist at the various interfaces within the SDH multiplexer, the signals are designed to float with respect to each other. This is known as *frequency justification* and is controlled using *pointers*.

Each pointer is located in two bytes, separated by a further two bytes (at the AU-4 level), or a complete frame (at the TU-12 level) with the construction as shown in Fig. 8.5. The pointer contains a value which determines the offset to the start of frame for the next level in the

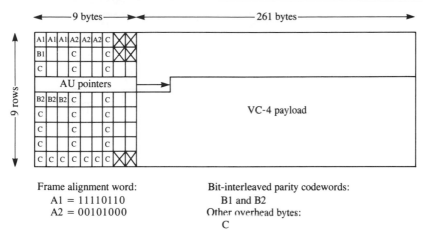

Figure 8.6 SDH pointer operation

hierarchy. For example, the two-dimensional representation of the
STM-1 frame is shown in Fig. 8.6, where transmission occurs from left
to right across each row in turn, and illustrates how the AU-4
pointer identifies the start of the VC-4 frame within its payload.
Frequency justification is achieved by incrementing or decrementing
the value of the pointer as required. This operation is performed by
inverting five bits of the 10-bit pointer word, either the I or D bits as
identified in Fig. 8.5. The demultiplexer decodes the pointer bits
using a majority decision algorithm in a similar way to the PDH.
However, this process could be more vulnerable to error activity
since most of the pointer bits reside in a single byte, unlike the PDH
where the justification service digits are distributed throughout the
frame to improve their error immunity. A theoretical study of the
alternative processing schemes has been performed by Deslandes *et al.*
(1991), with work also performed on a computer model of the error
propagation mechanisms.

Another significant advantage of the SDH is the way in which the
STM-n (n = 1, 4, 16) frame is constructed for transmission. In the case
of the PDH, each multiplexed level is further multiplexed as we
progress up the hierarchy, with access being lost to all overhead bits
until demultiplexed at the distant end. This cascading of the different
frame structures is a major contributor to the poor error propagation
performance identified within the PDH (Sec. 8.4.1). However, as
explained in Chapter 6, the SDH multiplexes the required number of
AUGs together before adding the overheads, such as FAWs, and
transmitting across the network. This removes the need for any
cascaded frame structures. For example, transmission at 622.08 Mbit/s
(STM-4) involves four AUGs to construct the frame and not four STM-
1 frames.

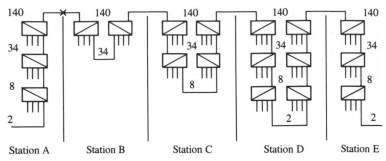

Figure 8.7 Hypothetical (European) PDH circuit routeing

8.5 Network routeing

8.5.1 Plesiochronous digital hierarchy

The services currently carried across digital networks fall into two main categories: interconnection of digital exchanges/switches/offices, and provision of private circuits. Both of these services are routed through a network by means of primary rate bearers, operating at 2.048 Mbit/s in Europe and 1.544 Mbit/s in the USA and Japan. However, transmission normally occurs at a higher transport rate of 140 or 565 Mbit/s for the European hierarchy, and 45 or 274 Mbit/s for the USA/Japan hierarchies. In both cases the ideal route for a circuit involves a single step up and down the digital hierarchy. However, in practice, the distances involved with the majority of routes demand the use of a number of higher order line systems, with routeing occurring at different levels within the hierarchy.

A hypothetical example of a 2 Mbit/s route is given in Fig. 8.7, which illustrates how a circuit may be routed over a number of bearers at different levels within the European hierarchy. This has the effect of disseminating the loss-of-alignment condition throughout the network, and results in alarms appearing at locations which are remote from the cause of the problem. For example, an error event occurring at point X (Fig. 8.7), resulting in a complete collapse of the hierarchy, will yield 140–34, 34–8 and 8–2 Mbit/s loss of alignment alarms at stations B, C and D respectively—this situation will be repeated for all 2 Mbit/s bearers carried on the 140 Mbit/s line system.

8.5.2 Synchronous digital hierarchy

Routeing through the PDH proves difficult because of the lack of access to the 2 Mbit/s tributary from the higher order multiplexed frames, resulting in the need to demultiplex the transport rate completely before routeing can occur.

The SDH overcomes these problems because of the ability to demultiplex a single 2 Mbit/s tributary directly from any transport level (STMn). This has led to the development of add–drop multiplexers to provide a flexible access to circuits and enable novel network topologies to be developed, as discussed in Chapter 6. The configuration of such multiplexers should eventually be possible under the control of network management software, greatly improving the circuit set-up time. Furthermore, the routeing of 2 Mbit/s circuits from one transport link to another will be possible using large cross-connect switches, also operating under software control.

8.6 Error performance monitoring

For many applications it is considered highly desirable to monitor the occurrence of transmission errors while a system is in service. This enables problems to be seen quickly and appropriate action to be taken, such as selecting an alternative link using automatic protection switching.

Errors are normally monitored by adding redundancy to the information signal. The added redundancy may primarily be used for other purposes (such as with line coding and frame alignment) or specifically for error detection (as with error control coding). These topics are dealt with in more detail elsewhere in this book; here we simply note a few points.

8.6.1 Line coding

Codes such as HDB3, 7B8B and the Manchester code can provide a monitoring capability on a per link basis. These codes have well-defined coding rules which can be verified either at a regenerator or a subsequent link-terminating point. Once any requisite codeword synchronism has been obtained, any violations of the coding rules can be attributed to transmission errors and so the error performance of the link can be estimated.

8.6.2 Frame alignment

One problem when monitoring line code violations occurs when a code conversion takes place, either to a new coding scheme for onward transmission, or to binary data for a multiplexing/demultiplexing operation. At this time any errors in the codewords will be incorporated into the data, and the new format will represent correct codewords derived from errored data. Even if every link in a circuit were monitored for code violations, and some form of correlation were performed to determine the end-to-end circuit performance, it would still not be possible to monitor errors introduced at the binary level, such as by multiplexing equipment. The only way to overcome such a limitation is to monitor known data within the frame that will be subjected to the same error activity as the tributary data.

One source of suitable data available within the frame is the frame alignment word. Once correct alignment is assured, any variation from the known FAW can again be attributed to transmission errors. As with line codes, care has to be taken to distinguish between word misalignment conditions and transmission errors. The decision algorithms are often quite complicated but nevertheless can give a reasonable estimate of transmission error performance. Frame alignment words usually remain associated with the transmitted information over more than a single link and so can offer the opportunity for error monitoring over a greater portion of a connection. For example, the 2 Mbit/s FAW will be carried through the PDH (or SDH) network until the 2 Mbit/s signal itself is demultiplexed. Further methods of embedded monitoring are considered in the next section.

8.6.3 Embedded monitoring

Cyclic redundancy check

A further method is available for monitoring end-to-end performance through the PDH which gives considerable improvements over the use of the FAW. This involves passing a number of frames through a *cyclic redundancy check* (CRC) circuit, and locating the generated remainder in spare bits of the frame overhead. For example, the *extended superframe format* of the North American T1 (1.544 Mbit/s) frame structure has been available for some time, and involves putting 24 × T1 frames (4656 bits) through a CRC generator to produce a six-bit codeword (CRC6) to be transmitted in the following 24-frame sequence (CCITT, 1988a). This provides error performance monitoring and also enhances the reliability of the receiver's framing algorithm. The European PDH has a similar system where 8 × 2 Mbit/s frames produce a four-bit codeword (CRC4) which is carried in the nonframe alignment word of timeslot 0 (CCITT, 1988a). A useful discussion on the operation of CRC4 can be found in McLintock and Harrison (1988).

Bit-interleaved parity

The SDH has been developed with many bits available at various levels within the hierarchy for embedded monitoring of the end-to-end performance. This involves a *bit-interleaved parity-X* (BIP-X) code whereby the transmitting equipment monitors a specified portion of the signal, and sets the code to provide even parity. For example, the performance between regenerator stages can be monitored using the B1 byte (Fig. 8.6) which employs a BIP-8 code. Each bit in the B1 byte produces an even parity on the corresponding bit of every byte in the previous frame. Other code sizes that exist include BIP-2 and BIP-$n \times 24$ (where n corresponds to the level of the STMn frame) and are described in CCITT (1990a,b).

8.6.4 Error control coding

It may not always be viable to provide an end-to-end level of error performance which is compatible with the most demanding digital service likely to be supported. In such circumstances error control coding (detection and then correction) over certain problem links or end-to-end over the whole connection may be used, as discussed in Chapter 9. This is the most direct method of error performance monitoring as the redundancy will usually have been specifically added, and so optimized, for the detection of errors. Of course certain error combinations may go undetected, or only be partially detected, and this can lead to erroneous action in a subsequent error corrector. However, this problem can usually be reduced to an acceptable level by using coding which is appropriate to the error characteristics of the particular transmission link. It then becomes possible to ensure that transmission errors in blocks of data, for example, are monitored to a high degree of confidence.

8.7 Error propagation through a network

It has been suspected for some time that lightning can induce error events on digital transmission systems. This has been confirmed by monitoring alarms within BT's digital network and correlating them with lightning events occurring on the UK mainland. For example, close correlation between lightning activity and error events in a digital network was found on a 140 Mbit/s coaxial line system between Bristol and Taunton. Monitoring equipment within the digital network recorded over 20 alarms which correlated with a lightning event in the vicinity of a regenerator. The alarms included a report of error activity on the line system and the lower order demultiplexers associated with the circuit losing alignment at Taunton, Exeter (30 miles beyond Taunton) and at Plymouth (40 miles beyond Exeter). A breakdown of the demultiplexer alarms generated by this example is given in Table 8.3.

Table 8.3 Demultiplexer alarms resulting from a lightning event

	140–34	34–8	8–2
Taunton	1	1	3
Exeter		3	11
Plymouth			1

8.8 Conclusions

This chapter has provided an introduction to the methods available for measuring the error performance of digital transmission systems. We have discussed the limitations of measuring the bit–error ratio alone and highlighted the importance of adding redundancy into the frame structures to provide an end-to-end performance monitoring capability. The opportunity for multiplexing equipments to extend the error activity through their frame alignment and justification procedures has also been demonstrated. In particular, the justification mechanism within the PDH makes the digital network vulnerable to short error bursts with a high bit–error probability, illustrated by the example of a 15 µs burst giving 1.2 ms of error activity at 64 kbit/s. The strict hierarchical nature of the current network has been shown to cause routeing difficulties, and also leads to error events mushrooming through the network and generating alarms, some of which can be at a considerable distance away from the disturbance.

References

Boehm, R. J., Ching, Y.-C. and Sherman, R. C. (1985) 'SONET (Synchronous Optical Network)', *IEEE Globecom. Conference*, vol. 13, pp. 46.8.1–8.

CCIR (1986) 'Digital television: transmission impairments and methods of protection', Report 967–1.

CCITT (1972a) 'G.736, Characteristics of a synchronous digital multiplex equipment operating at 2048 kbit/s', in *Blue Book*.

CCITT (1972b), 'G.742, Second order digital multiplex equipment operating at 8448 kbit/s and using positive justification', in *Blue Book*.

CCITT (1976) 'G.751, Digital multiplex equipments operating at the third order bit rate of 34 368 kbit/s and the fourth order bit rate of 139 264 kbit/s and using positive justification', in *Blue Book*.

CCITT (1988a) 'G.704, Synchronous frame structures used at primary and secondary hierarchical levels', in *Blue Book*.

CCITT (1988b) 'G.821, Error performance of an international digital connection forming part of an integrated services digital network', in *Blue Book*.

CCITT (1988c) 'M.550, Performance limits for bringing into service and maintenance of digital paths, sections and line sections', in *Blue Book*.

CCITT (1990a) 'G.707, Synchronous digital hierarchy bit rates', approved version SG XVIII.
CCITT (1990b) 'G.708, Network node interface for the synchronous digital hierarchy', approved version SG XVII.
CCITT (1990c) 'G.709, Synchronous multiplexing structure', approved version SG XVII.
Deslandes, J. E. (1991) 'Error propagation through digital demultiplexers', Ph.D. dissertation, University of Essex.
Deslandes, J. E., Cochrane, P. and Jones, E. V. (1990a) 'European multiplex hierarchy—mechanisms for error extension', *IEEE Globecom. Conference,* vol. 12, pp. 701.5.1–5, December.
Deslandes, J. E., Cochrane, P. and Jones, E. V. (1990b) 'Error propagation through the European demultiplexer hierarchy—an analysis', *IEE Electronics Letters*, vol. 26, no. 18, 1469–70.
Deslandes, J. E., Jones, E. V. and Cochrane, P. (1991) 'Analysis of SDH pointer corruption due to error activity', *IEE Electronics Letters*, vol. 27, no. 5, 453–5.
Kanal, L. N and Sastry, A. R. K. (1978) 'Models for channels with memory and their applications to error control', *Proceedings IEEE*, vol. 66, 724–44.
Knowles, M. D. and Drukarev, A. I. (1988) 'Bit error rate estimation for channels with memory', *IEEE Transactions on Communications*, vol. 36, 767–9.
Kubat, P. and Bollen, R. E. (1989) 'A digital circuit performance analysis for tandem burst-error links in an ISDN environment', *IEEE Transactions on Communications*, vol. 37, 1071–6.
McLintock, R. W. and Harrison, N (1988) 'Introduction of a cyclic redundancy check procedure into the 2048 kbit/s basic frame structure', *British Telecommunications Engineering*, vol. 6, part 4, 218–24.
McLintock, R. W. and Kearsey, B. N. (1984) 'Error performance objectives for digital networks', *The Radio and Electronic Engineer*, vol. 54, 79–85. (A later version which includes changes in CCITT recommendations is to be found in *British Telecommunications Engineering*, vol. 3, 92–98.)
Owen, F. F. E. (1982) *PCM and Digital Transmission Systems*, Chapter 7, McGraw-Hill, New York.
Takahashi, K. (1988) 'Transmission quality of evolving telephone services', *IEEE Communications Magazine*, October, vol. 26, no. 10, 24–35.
Yamamoto, Y. and Wright, T. (1989) 'Error performance in evolving digital networks including ISDNs', *IEEE Communications Magazine*, vol. 27, no. 4, 12–18.

9 Error control coding

KEN CATTERMOLE

9.1 General ideas

9.1.1 Purpose

Much of communication theory and practice concerns the conveyance of signals through imperfect channels. The signals must be designed to resist the channel impairments as far as possible; the receiver must be designed to recover the signals as accurately as possible. Significant impairments occurring in practice may be classified as:

1 *Additive noise*
   (a) Gaussian noise, white or coloured, with stationary statistics
   (b) Impulsive noise, of diverse origin in different physical channels, not always stationary or easy to characterize

2 *Channel perturbations*
   (a) Fading in radio channels
   (b) Sync slip in digital channels
   (c) Breaks in transmission, of diverse origin

The likely effects on digital signals may be classified as:

1 *Uniformly random errors* Digital errors occurring individually and independently, with approximately uniform probability density, primarily due to noise (often called just 'random errors').

2 *Burst errors* Digital errors grouped in clusters, mainly the result of a combination of noise and channel perturbations.

3 *Erasures* Irregular intervals when it is known that no reliable signal can be detected, because of severe channel perturbation.

A suitable choice of digital code and decoding technique can substantially reduce the probability of error, or even eliminate errors entirely, so long as the impairments are no worse than a specified condition for which the system is designed. All such systems may, however, be overborne by impairments which are excessive in magnitude or adverse in pattern. The system designer's task is to guard against the more probable types of error, accepting that (one hopes with very low probability) the system may fail.

9.1.2 Principle

All error control procedures rely on the principle of redundancy, that is to say, the signalling rate of the channel must be greater than the rate of information conveyed. The additional signalling capacity enables constraints to be imposed on valid signals transmitted: if the constraints are violated by the received signal, it is known that an error has occurred, and it may be possible to deduce the nature of the error.

The simplest form of redundancy is repetition. Suppose an elementary signal is repeated n times. If the receiver detects n identical signals, it is a reasonable presumption that these are correct. If the repeated signals are not identical at the receiver, clearly an error has occurred. If $n > 2$, it may be possible to establish by majority vote which signal out of the possible repertoire is most likely to be correct.

More generally, only a subset of possible digit sequences are valid signals: errors usually produce sequences which are recognizable as invalid. It is important that the subset of valid sequences can be specified by some logical rule, so that the receiver may use an algorithm for testing sequences and making inferences from the form of discrepancy observed. Secs. 9.2–9.4 describe several useful encoding and decoding procedures. Very commonly, a coded sequence is formed by taking a given data sequence and extending it by appending or interleaving a further sequence of check digits generated from the data digits by means of an algebraic procedure.

The following remarks, though based on a special case, apply to all the types of code which we shall consider. It is clear that if an elementary signal is repeated, the following outcomes are possible:

- Signal received with no detectable error (all n repetitions alike).
- Error detected but a signal value can be inferred (a majority of the n repetitions alike, a minority of discrepancies).
- Error detected but a signal value can not be inferred (no clear majority, most obviously if n is even and the repetitions divide equally between two alternatives).

In the first two cases it is customary to assume that the inferred signal is correct and to say that the errors in the second case have been corrected. However, one must be aware that the inferred signal may possibly be wrong. Multiple errors can occur and can give a false indication: perhaps a majority, or even all n repetitions, suffer the same error. Such errors cannot be detected by examination of the received signal sequence. It is a necessary assumption in operating an error control system that undetectable errors have a much lower probability of occurrence than detectable errors. The design must be such that this assumption is normally true.

9.1.3 Procedures

A digital signal comprises a sequence of symbols which either are binary or are translated from and to binary sequences for the purposes of encoding and decoding. We shall assume unless otherwise stated that the symbols are individually detected, that effectively the digits of a binary sequence are individually detected, and that a digit error is a binary error, namely a 1 is converted to a 0 or vice versa. The encoder and decoder are acting on binary digit streams.

There are three procedures in common use, separately or in combination, namely:

1   *Error detection and block repetition* (often referred to as ARQ—automatic request for retransmission)   The source data is partitioned into blocks of k bits, converted by the encoder into blocks of $n(>k)$ bits with enough checks to enable the decoder to detect errors of the more probable kinds. This system requires a return channel, over which the receiver sends an acknowledgement signal to the data source; various systems use positive acknowledgement (receipt of a correct block), negative acknowledgement (receipt of an erroneous block) or both. If an error is detected, the source responds to a negative acknowledgement, or the absence of a positive acknowledgement, by repeating the block containing the error.

2   *Forward error correction (FEC) using block codes*   Source data is partitioned into blocks of k bits, converted by the encoder into blocks of $n(>k)$ bits with enough checks to enable the decoder to correct errors of the more probable kinds. Error-correcting codes have more redundancy than error-detecting codes, and the decoding algorithms are much more complex.

3   *Convolutional coding*   Here the encoder operates not on disjoint blocks, but on a running block of bits held in a shift register, generating a sequence of higher rate. This procedure is normally used for FEC, but the correcting capabilities are not so clear cut as with block codes. Probabilistic decoding, approximating maximum likelihood, is generally used.

Other procedures which may be utilized, especially on rather 'difficult' media, include: provision for handling erasures, interleaving of blocks to spread burst errors, and concatenation of two different coding methods.

Section 9.2 presents some general ideas and introduces simple block codes. Section 9.3 lays a foundation for the design and analysis of the more complex codes widely used in practice. Section 9.4 gives an introductory account of convolutional coding and the associated Viterbi decoding procedure.

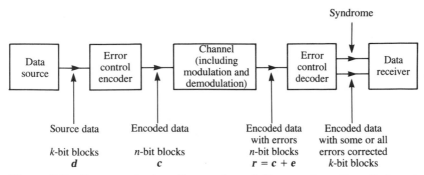

Figure 9.1 Error control coding system. (**d** is a vector of length *k* and **c**, **r** and **e** are vectors of length *n*. All have binary components)

9.2 Simple block codes

9.2.1 The geometric picture

In a block code, the source data is partitioned into blocks of *k* bits. Each block is translated by the encoder into a block of *n* bits ($n > k$) for transmission through the imperfect medium (Fig. 9.1). The code is conventionally known as an (*n,k*) code. Usually, the transmitted block contains the *k* data bits unchanged, together with ($n − k$) check bits generated from the data bits according to a defined algorithm. At the receiver, the decoder recovers the data and generates a *syndrome*, which is a coded indication of the error state utilized in the process of correction.

The error control capabilities of the code depend on a key property which can be specified algebraically or geometrically. The *Hamming distance* [1] between two codewords is defined as the number of bit positions in which the words differ. For example:

$$\text{Distance } (0110, 1010) = 2 \qquad (9.1)$$

We can consider a codeword as a vector in *n*-dimensional space, and (at least for small *n*) draw a diagram in which codewords appear as vertices of an *n*-dimensional hypercube. Figure 9.2 shows two examples: (a) with $n = 3$ and all possible three-bit vectors included; (b) with $n = 4$ and the set of eight vectors

$$
\begin{array}{ll}
0000 & 1001 \\
0011 & 1010 \\
0101 & 1100 \\
0110 & 1111
\end{array}
\qquad (9.2)
$$

---

[1] Named after Richard W. Hamming, a pioneer of error-control coding. We define and utilize the distance concept only for binary symbols, though it can be generalized to deal with higher radices.

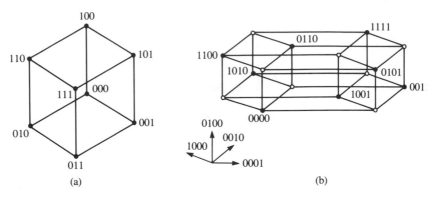

Figure 9.2 Geometry of codes

distinguished as codewords. This (4,3) code includes the two words which appear in Eq. (9.1). To join their representative points by a path along the edges of the hypercube, at least two edges must be traversed, corresponding to the stated distance, 2. Similarly, the distance between any two words of the set is at least 2. The other eight vertices in the diagram correspond to binary vectors, such as 0001, which are not in the codeword set.

The distance d of a code is defined as the minimum distance between any pair of codewords. If $d = 1$, no error control is possible: an error pattern can convert a valid word into another valid word. If $d = 2$, as in Fig. 9.2(b), then any single-bit error converts a valid word into an invalid word, so the occurrence of a single error (more generally, an odd number of errors) can be detected by a decoder. It is not possible to identify and correct the error, since the invalid word is equally distant from several valid words, and there is no way of choosing between them.

Now consider a code with $d = 3$. For a simple example, suppose that in Fig. 9.2(a) we designate as codewords only the diagonally opposite vertices 000 and 111; this is a (3,1) code amounting to triple repetition. Any single error converts a valid source word into an invalid received word which is at distance 1 from the true source word but distance 2 from the alternative valid word. On the basic assumption that single errors are more probable than double errors, the source word may be inferred and the single error corrected. Each valid word is surrounded by a decoding sphere comprising all vectors nearer to the given word than to any other valid word. In this example the decoding spheres are

$$[000, 001, 010, 100]$$
$$[111, 110, 101, 011]$$

(9.3)

which may be interpreted as binary 0 and 1 respectively.

Continuing with a code of distance 4, suppose that in Fig. 9.2(b) we

designate as codewords only the diagonally opposite vertices 0000 and
1111; this is a (4,1) code amounting to quadruple repetition. Around
each codeword is a decoding sphere including points at distance 1, but
between the spheres are points at distance 2 from both codewords. A
double bit error gives a received vector in one of the interstitial locations:
an error has obviously occurred, but cannot be corrected.

The properties exhibited in these examples generalize in a fairly
obvious way to any code of given distance d, as follows. With suitable
decoding procedures, it is possible to:

Detect a-bit errors if $a < d$	(9.4)
Correct b-bit errors if $2b < d$	(9.5)
Correct b-bit errors and detect a ($> b$) bit errors if $a + b < d$	(9.6)

The statements based on repetition (in Sec. 9.1.2) are special cases of
(9.4) and (9.5), since n-fold repetition is simply an $(n,1)$ code of distance
$d = n$. A major topic in coding theory is to design more efficient codes
with useful values of d. The remainder of this chapter introduces the
concepts needed to accomplish this task.

9.2.2 Parity checks

An important property of a binary codeword is its parity. This has two
values, odd or even, according to the number of bits which have the
value 1. Equivalently, we may define the parity of a word as the modulo
2 sum of its bits:

$$\text{Parity } (d_1, d_2, \ldots, d_k) = d_1 + d_2 + \cdots + d_k \quad (\text{mod } 2) \quad (9.7)$$

which has the value 0 for even and 1 for odd parity. Each of the words
in expression (9.2) or Fig. 9.2(b) has even parity, there being either 0, 2,
or 4 bits with the value 1.

Any binary word may be translated into a longer word of defined
parity by appending a suitably chosen parity bit. If:

$$c_i = d_i \quad (i = 1, 2, \ldots, k)$$
$$c_{k+1} = d_1 + d_2 + \cdots + d_k \quad (\text{mod } 2)$$

then:

$$\text{Parity } (c_1 c_2 \ldots c_k c_{k+1}) = 0 \quad (9.8)$$

This translation is the simplest possible error-detecting code, with
parameters $(k + 1, k)$. Any single bit error in an even-parity word
converts it to odd parity. So the encoder appends a parity bit; the
decoder takes a sum modulo 2 over all bits, this sum being a syndrome
whose value is 1 for detected error and 0 for no detected error.

From a geometric viewpoint, the distance between two words of even
parity must itself be even; we have already verified this in the example

d_{11}	d_{12}	$\ldots$	d_{1s}	$c_{1(s+1)}$
d_{21}	d_{22}	$\ldots$	d_{2s}	$c_{2(s+1)}$
.	.		.	.
.	.		.	.
.	.		.	.
d_{r1}	d_{r2}	$\ldots$	d_{rs}	$c_{r(s+1)}$
$c_{(r+1)1}$	$c_{(r+1)2}$	$\ldots$	$c_{(r+1)s}$	$c_{(r+1)(s+1)}$

Figure 9.3   Array codes

of Fig. 9.2(b). So the minimum nonzero distance is 2; by (9.4) the code can detect single errors, but by (9.5) it cannot correct any error.

More generally, any odd number of errors may be detected by means of a parity check. It may be possible to turn this property to advantage. Suppose we have a code of distance 3, thereby capable of detecting double errors, such as the Hamming code of Sec. 9.2.4. Clearly, the codewords are not all of similar parity. By appending a parity bit, the minimum distance is increased to 4, and all errors up to triple may be detected.

So far we have described only a single parity check, taken over all bits of a codeword. A large class of error-control codes, including those described below, make use of multiple parity checks. Several check bits are appended, each of which controls the parity of a distinct subset of the bits of the codeword. The error detecting and correcting properties depend on the number of checks and the choice of subsets.

9.2.3 Array codes

A simple way of defining subsets for parity checking is to consider the bits as a rectangular array, whose rows and columns constitute natural subsets. Figure 9.3 shows $k = rs$ data bits arranged as r rows, each of s bits. To each row is appended an $(s + 1)$th bit chosen to make the row parity even. To each column is appended an $(r + 1)$th bit chosen to make the column parity even. Finally, the bit in position $(r + 1, s + 1)$ is an overall parity bit.

Changing any one bit alters three parities—one row, one column and overall. This enables any single error to be located—it is at the intersection of the row and column whose parities are upset. A double error changes two parities if both errors are in a common row or column, four parities otherwise. It can be detected but not corrected, since in each case two or more error patterns can yield the same syndrome. A triple error changes overall parity and various row and column parities, again detectable but not correctable. A quadruple error may fail to be detected: consider four bit errors at the corners of a rectangle, which violate no parities whatsoever. We conclude that the code has distance 4, and can be used either to correct one error and detect two, or to detect three errors.

Codes of this class are by no means the most powerful or efficient, but as well as having some practical uses, they demonstrate several important points. Firstly, they show that the subset-parity technique can yield distances of 3 or 4 in a simple and general way, much more efficiently than plain repetition, which is the most obvious form of redundancy.

Secondly, they share with the most efficient codes the useful property that errors of higher weight (i.e., number of bits in error) than the design value may be detected with high probability. Detectability depends not only on weight of error but also on pattern. We have remarked that a specific set of quadruple errors are undetectable, but these are a small minority. With blocks of useful practical size, most quadruple errors are detectable. For example, in a (64,49) code only 0.0012 of quadruple errors are undetectable, and in a (256,225) code only 0.000082.[2] All errors of odd weight are detectable (they violate overall parity) and so are a large majority of higher weight errors of even weight.

Thirdly, we recall the distinction made in Sec. 9.1.1 between individually random errors and burst errors. The error-detecting criterion (9.4) applies to errors of arbitrary pattern and so to random errors. Most error-detecting codes will detect errors of higher weight if these are clustered into a burst. To illustrate this property for an array code, we must define the way in which the bits of the array are turned into a sequence for transmission. An obvious method is to scan by rows (or columns—there is no difference of principle). We then have two sets of parity checks: the row parities over sets of adjacent bits; the column parities over interleaved sets at regular intervals $(s + 1)$. Now consider the effect of a burst of errors of any length up to that of a row. It may or may not upset row parity; it is certain to upset several column parities, and so be detected. The detectable burst length may be, and in practice is, much longer than the general detection capability of 3. This illustrates the benefit of interleaving, which is often used over channels liable to correlated effects such as radio fading. Clustered errors are dispersed among several independently checked subsets and thereby rendered more readily detectable.

Fourthly, the array code is the simplest example of the principle of concatenation. In concatenated encoding, source data is applied to an outer encoder, then to an interleaver, then to an inner encoder, with complementary operations at the decoding terminal. In this case, both encoders are simple parity checks. In the general case, they may be more

---

[2] This follows from a simple combinatorial argument. Try it as an exercise.

powerful encoders, and the combination may be very effective especially against burst errors.

9.2.4 Hamming codes

This class of codes, one of the first to be devised, has distance 3 (easily extended to 4) and so is single-error-correcting. Unlike the array code, it uses a minimal number of check bits to attain this performance.

The minimal number of checks may be found as follows. If we wish to correct any single error in an n-bit word, then there must be at least $(n + 1)$ distinct syndromes—one for each bit position which may be in error, and one to denote the absence of detected error. So the number of check bits, $c = n - k$, must be such that:

$$2^c \geqslant n + 1 \tag{9.9}$$

The maximum values of n and k consistent with a given c are therefore:

$$n = 2^c - 1 \tag{9.10}$$
$$k = 2^c - c - 1 \tag{9.11}$$

suggesting the possibility of distance 3 codes with parameters (3,1): (7,4) : (15,11) : (31,26) : (63,57), and so on.

There are several ways of choosing parity check subsets to attain these capacities, but perhaps the most readily intelligible is Hamming's original version. The syndrome is a c-bit binary number. Let the zero syndrome denote no detected error, and let each nonzero syndrome be the binary representation of the (single) erroneous bit position.

For example, with the (7,4) code we need three subsets, say x, y, and z respectively, which include bits chosen from positions 1–7 as in the following table (1 denotes inclusion, 0 exclusion):

Bit position	1	2	3	4	5	6	7	
Subset x	0	0	0	1	1	1	1	
Subset y	0	1	1	0	0	1	1	(9.12)
Subset z	1	0	1	0	1	0	1	

In encoding to this rule, positions 3,5,6,7 are allotted to data bits: positions 1,2,4 (in general, powers of 2) to check bits, each of which is in just one subset and can be chosen to give subset parity 0. At the decoder, if there is no error then parity is 0 in each subset. If there is an error in any one location i, then the syndrome comprising the three parities xyz gives the binary representation of i. For example, the four-bit data word 1001 is encoded as the seven-bit word 0011001. If as a result of transmission errors the received word is 0011011, the syndrome is 110, indicating that the error is in the sixth bit which can then be corrected.

The extension to longer codes with parameters given by Eqs. (9.10) and (9.11) is obvious. If blocks of intermediate length are required, take the check structure of the next n value above and drop the requisite number of data bits. Clearly, check bits cannot be dropped since the inequality (9.9) must always be satisfied.

Other valid choices of subset may be made by permuting the columns of an array such as (9.12), for instance to place the data bits in contiguous positions. As we shall see in Sec. 9.3, this is commonly done in practice, being convenient for the most useful encoding and decoding mechanisms.

Since the code is single-error-correcting, its distance d must be at least 3. It is easily shown to be not greater than 3, by the following argument. The data word comprising k 0s translates into the codeword comprising n 0s. A data word with a single 1 in a location whose binary number contains two 1s translates into a codeword with three 1s. So there are codeword pairs of distance 3, and the code distance is 3. By the argument of Sec. 9.2.2, appending an overall parity bit increases the distance to 4. So there are triple-error-detecting codes with parameters (4,1): (8,4): (16,11): (32,26), and so on.

The ability of the Hamming codes to detect burst errors is not apparent from this approach, but by using the cyclic structure developed in Sec. 9.3 it may be shown that a suitable version can detect all burst errors up to length $n - k$. In practical cases such as $n = 256$, this is significantly longer than the general detection capability of 2 or 3.

We remarked in Sec. 9.2.3 that usually errors of higher weight than the design value may be detected with high probability. An error pattern of weight d is undetectable if and only if it causes one codeword to be converted into another codeword—a majority of patterns do not have this property. A combinatorial calculation shows that for the basic Hamming (n,k) distance 3 code, or the extended $(n + 1,k)$ distance 4 code, the proportion of undetectable d-fold errors is $1/(n - 2)$. For example, in a (64,57) code only 0.016 of quadruple errors are undetectable, and in a (256,247) code only 0.0040. It may also be shown that bursts longer than the critical length $(n - k)$ have a high probability of detection.

9.2.5 Probabilistic analysis of code performance

The occurrence of errors in a channel is a random process, and if the statistics of the process are known then in principle the statistics of the decoder output can be estimated. The simplest case is that of independently random bit errors of equal probability p. The number of errors in a block then has a binomial distribution, the probability of an error of weight r being:

$$p_r = \binom{n}{r} p^r (1 - p)^{n-r} = \binom{n}{r} p^r + O(p^{r+1}) \qquad (9.13)^3$$

the approximation applying for small values of p such as might be expected in a channel of very good quality. With small error rates, this distribution decreases monotonically and fairly rapidly with increasing r. This is the justification for designing codes which can detect or correct errors up to a certain weight.

Consider a code which can detect a errors. With the probability distribution (9.13) the probability of the three significant events can be easily bounded:

Prob(no error)	$= p_0$	(9.14)
Prob(detected error)	$> p_1 + p_2 + \cdots + p_a = O(p)$	(9.15)
Prob(undetected error)	$< p_{a+1} + p_{a+2} + \cdots = O(p^{a+1})$	(9.16)

The inequalities in (9.15) and (9.16) are explained by the fact that all errors of weight $1, 2, \ldots, a$ are detected plus some of the errors of greater weight. We can be more precise if we know the weight distribution of undetectable errors, which in the case of group codes (a category defined in Sec. 9.3 and including all the codes discussed here) can be shown to be the same as the weight distribution of the code. Denote the number of words of weight r by A_r, then:

$$\text{Prob(undetected error)} = A_{a+1} p^{a+1} + O(p^{a+2}) \qquad (9.17)$$

For the basic Hamming code, we have $a = 2$ and:

$$A_3 = \frac{1}{n-2} \binom{n}{3} \qquad (9.18)$$

whence:

$$\text{Prob(undetected error)} = \frac{1}{6} n(n-1) p^3 + O(p^4) \qquad (9.19)$$

For the extended Hamming code, $a = 3$ and:

$$A_4 = \frac{1}{n-2} \binom{n+1}{4} \qquad (9.20)$$

whence:

$$\text{Prob(undetected error)} = \frac{1}{24} n(n^2 - 1) p^4 + O(p^5) \qquad (9.21)$$

Now consider a code which can correct b errors. The significant

---

[3] The notation $O(p^{r+1})$ signifies a remainder which, when expanded in powers of p, has a leading term of degree not less than $r + 1$, and so becomes relatively small as p decreases.

events are: no error; corrected error (received vector falls in the correct decoding sphere); decoding error (received vector falls in a wrong decoding sphere); and decoding failure (received vector falls in the interstices between decoding spheres). The probabilities easily calculated are:

Prob(no error) $\qquad\qquad = p_0$ $\qquad\qquad\qquad\qquad$ (9.22)
Prob(corrected error) $\qquad = p_1 + p_2 + \cdots + p_b = O(p)$ $\qquad$ (9.23)
Prob(decoding error or failure) $= p_{b+1} + p_{b+2} + \cdots = O(p^{b+1})$ (9.24)

Notice that in (9.23) and (9.24)—unlike the error-detecting case—we have written equalities rather than inequalities. Whereas an error-detecting system usually detects most errors of weight $> a$, an error-correcting system does not normally correct errors of weight $> b$. Estimation of the probabilities stated is straightforward, but partitioning of decoding error and decoding failure may be more complex.

The basic Hamming (n,k) code with $b = 1$ is simple to analyse because there is no space between the decoding spheres. So:

$$\text{Prob(decoding error)} = p_2 + p_3 + \cdots$$
$$= \tfrac{1}{2}n(n - 1)p^2 + O(p^3) \qquad (9.25)$$

The extended Hamming $(n + 1,k)$ code has $b = 1$ but can also detect $a = 2$ errors: a similar expression gives the sum of error and failure probabilities. In this case, we can easily partition the decoding errors and failures, since all words of weight 3 fall in decoding spheres and all words of weight 2 in interstices. Consequently:

$$\text{Prob(decoding failure)} = \binom{n + 1}{2}p^2 + O(p^3) \qquad (9.26)$$

$$\text{Prob(decoding error)} \;\; = \binom{n + 1}{3}p^3 + O(p^4) \qquad (9.27)$$

We conclude with a numerical example. For $n = 63, p = 0.001$, the probabilities (9.19) and (9.21) are about 6.5×10^{-7} and 1.0×10^{-8}, respectively. Since the error probability is about 0.06, it is clear that most errors are detected, and the residual probability of undetected error is very low. However, the probabilities (9.25) and (9.26) are about 0.0018, which is far from trivial in most applications. This comparison explains why, although ARQ systems may well employ a fairly simple code such as extended Hamming, FEC systems employ more powerful codes. Such codes can be designed only with the aid of an algebraic theory, introduced in Sec. 9.3.

9.3 Algebraic structure of block codes

9.3.1 Introductory comment

The general ideas behind error control codes, and the design of single-error-correcting codes, are open to simple description. However, codes of distance 3 or 4, though adequate for some error-detecting applications, are not sufficiently powerful for practical error-correcting systems. Their limitations are:

- Nontrivial probability of decoding error or failure against moderate rates of random bit error,
- Drastic loss of performance as the bit error rate rises, because of complete failure to deal with errors of greater weight than the design value,
- Inability to cope with burst errors, which are common on some transmission media.

Practical error-correcting systems require, and use, codes with multiple-error-correcting capability, which implies distances from 5 upwards. These codes use the principle of subset parity, with a larger number of check bits. The checks must be chosen so that each possible error, of whatever weight within the design limit, gives a distinct syndrome. The design of codes with such properties is based on a structure whose explanation requires some background knowledge of linear algebra.

A short introductory treatment cannot give the algebraic background in much depth, or a comprehensive list of codes. The aim here is to introduce the algebraic concepts, and to show how one of the major classes of code—the BCH codes—relates to them.

For a fuller exposition of the basic mathematics, integrated with other topics in signal theory, and a proof of all properties cited here, see Cattermole (1986). For a comprehensive account of coding theory, there is a large specialist literature, of which Blahut (1983) is particularly recommended.

9.3.2 Groups, fields, and polynomials

There are many algebraic structures with the following common features: a set of mathematical objects; one or more operations which can be performed on objects from the set; and the rule that the result of an operation is a member of the set (this rule is known as closure). Familiar examples, which were historically the genesis of a wider theory, are provided by the number system. Here the objects of the set are numbers (real numbers, complex numbers, rational numbers, integers, etc.) and the operations are ordinary arithmetical addition, subtraction, multiplication, and division. The closure rule means that, for example, if we add two numbers (of any of the types mentioned) the result is another number of the same type. Not all operations are valid in all sets: for example, division of integers does not always yield an integer. A fundamental property of an algebraic structure is the list of valid operations.

Of the many known structures, we shall focus on two, groups and fields, which admit respectively one and two operations which can be thought of as addition/subtraction and multiplication/division.

The number of objects in the set is known as the order of the structure; it may be finite or infinite. The familiar number systems have infinite order; here we are more concerned with finite order. Groups of all orders exist, but fields exist only for a limited set of orders, namely the powers of prime numbers. Powers of two are, of course, important for the treatment of binary codes.

An example of a finite group is the set of integers:

$$Z_q = (0, 1, 2, \ldots, q - 1) \tag{9.28}$$

under the operation of addition modulo q.[4] This satisfies the closure rule: for any members a,b the sum $a + b$ (mod q) is a member of the set. There is an identity element, in this case 0, such that $a + 0 = a$. Each element a has an inverse $-a$, such that $a + (-a) = 0$.

Most binary codes, including all those mentioned here, have a group structure: the operation is bit-by-bit addition modulo 2 (Cattermole, 1986, p. 245). For example, the codeword set (9.2) is a group, with the all-zero word as identity: closure is illustrated by relationships such as $0011 + 1010 = 1001$. The parity of the sum of two words equals the sum of their parities (in this example, $0 + 0 = 0$).

A group G may have a subgroup H, namely a subset of elements forming a group under the same operation. For example, the group of integers modulo 10 has the subgroup (0, 2, 4, 6, 8), i.e., all the even integers in the group. The set of all four-bit binary words is a group of order 16: the codeword set (9.2) is a subgroup of order 8.

Suppose we take any element of G not in the subgroup H, and combine it with each member of H in turn. This generates a subset of G, of the same order as H, known as a coset. For example, with integers mod 10, add any odd number to the subgroup of even numbers—this generates the coset of all odd numbers. (Note that this is not another subgroup: it does not satisfy the closure rule.) With four-bit binary words, add any word of odd parity to the subgroup with even parity: this generates the coset of all words of odd parity.

To show the relevance of the group concept to coding, we cite two properties of group codes. Firstly, all n-bit words constitute a group G

---

[4] For formal definition of a group, and further examples, see Cattermole (1986).

of which the valid codewords are a subgroup H (Cattermole, 1986, p. 245). An error pattern is also an n-bit vector, which we can write with a 1 for each erroneous bit and a 0 for each correct bit. Referring to Fig. 9.1, let a codeword c be corrupted by an error pattern e, giving a received word $r = c + e$. The codewords c are drawn from a subgroup H. The error patterns which can be detected are not in this subgroup: the received words in the presence of any given error pattern constitute a coset. In an error-correcting code, distinct correctable errors generate distinct cosets, each with a characteristic syndrome, which is the basis for recognition by the decoder.

Secondly, a group has characteristic symmetry properties (Cattermole, 1986, pp. 221ff), some of which may be obvious from geometric pictures such as Fig. 9.2. In group codes, the distribution of distances from a nominated word to all other valid codewords is the same for any nominated codeword. In particular, it equals the distribution of distances from the zero word (which must be present in a code satisfying group properties), so it equals the distribution of codeword weights, i.e., the number of 1 bits in a word (Cattermole, 1986, p. 253). The minimum distance of a group code equals the minimum weight of its nonzero words.

A field differs from a group in that it has two operations, which we can think of as addition and multiplication (Cattermole, 1986, pp. 265ff). Taking numbers as an example, we require additive properties like those of a finite group of integers, together with similar properties under multiplication. The multiplicative identity is 1, since $a \times 1 = a$ for any a in the set. The multiplicative inverse is $1/a$, since $a(1/a) = 1$. Under addition, all numbers have an inverse; under multiplication, there is an exception, namely 0, which is the additive identity.

The set of numbers (9.28) mod q constitute a field under addition and multiplication if and only if q is prime (Cattermole, 1986, p. 258). It may be shown that division fails to work unless the divisor is relatively prime to q, and this can be the case for all divisors only if q is itself a prime number. The field of numbers modulo a prime p is called $GF(p)$, the notation signifying 'Galois field' after the mathematician Evariste Galois.

Consider, for example, the additive and multiplicative properties of $GF(5)$. Taking all numbers modulo 5, we have:

$$3 + 4 = 7 = 2 \bmod 5$$
$$2 \times 3 = 6 = 1 \bmod 5 \tag{9.29}$$

illustrating closure. The multiplicative group of a field has a further property: it is cyclic, meaning that all members may be generated by successive multiplication of one member, called a generator (Cattermole, 1986, p. 279). For example, 2 is a generator of $GF(5)$, since:

$$2^1 = 2, \quad 2^2 = 4, \quad 2^3 = 3, \quad 2^4 = 1, \quad 2^5 = 2 \qquad \text{mod } 5 \quad (9.30)$$

For any generator y of a field of order p:

$$y^p = y \qquad (9.31)$$

as in this example.

To apply the field concept to an entity with many components, such as a code, we have to find the appropriate form of the two operations, addition and multiplication. We have already seen that modulo 2 addition bit-by-bit works satisfactorily for codewords, or indeed for vectors in general. Multiplication requires a different representation, one which is familiar from elementary algebra: we take the elements as coefficients of a polynomial. The polynomial representation of an n-bit codeword is:

$$c(x) = c_0 + c_1 x + c_2 x^2 + \cdots + c_{n-1} x^{n-1} \qquad (9.32)$$

If the coefficients are chosen from a group of order q, then under addition the polynomials constitute a group of order q^n. For example, the set of codewords (9.2) has the polynomial representation:

$$
\begin{array}{lll}
0 & 1 + & x^3 \\
x^2 + x^3 & 1 + & x^2 \\
x + \quad x^3 & 1 + x & \\
x + x^2 & 1 + x + x^2 + x^3 &
\end{array}
\qquad (9.33)
$$

The multiplication of polynomials follows the usual rules, with one modification. Recall for a moment the finite field $GF(p)$ in its numerical form. All numbers are interpreted modulo a prime number p, and this is necessary to satisfy the field properties. Similarly, in a field $GF(p^n)$ of n-term polynomials with coefficients in $GF(p)$, we have to interpret products modulo a prime polynomial. That is to say, the remainder is taken after division of the apparent product by a polynomial which is prime in the sense that it has no factors with coefficients in the field $GF(p)$.

Consider the three-bit words represented by second-degree polynomials, with products expressed modulo $1 + x + x^3$. We have, for example:

$$
\begin{aligned}
(1 + x^2)(1 + x + x^2) &= 1 + x + x^3 + x^4 \\
&= (1 + x)(1 + x + x^3) + x + x^2 \\
&= x + x^2 \qquad \text{mod}(1 + x + x^3) \quad (9.34)
\end{aligned}
$$

giving another word of the set. The same would apply for any pair of words (Cattermole, 1986, p. 272). The multiplicative group is cyclic, one generator being x (Cattermole, 1986, p. 280). It is easily verified that:

$$x^8 = x \qquad \text{mod}(1 + x + x^3) \qquad (9.35)$$

and also that no lower power of x satisfies such an equation.

The polynomial representation enables us to define codes whose structure is not just a group but a field. This opens up some very useful techniques of design, analysis, and implementation, as we shall see in the remaining sections of this chapter.

9.3.3 Cyclic codes

We use the polynomial representation to define codes that have multiplicative as well as additive properties.

A cyclic code is a group code which has the further property that for any code word:

$$c = (c_0, c_1, \ldots, c_{n-1})$$

the cyclically shifted word:

$$c' = (c_{n-1}, c_0, c_1, \ldots, c_{n-2})$$

is also a codeword. In polynomial terms this implies that, if $c(x)$ is a codeword, then:

$$c'(x) = xc(x) \qquad \mod(x^n - 1) \tag{9.36}$$

is also a codeword. Taking this property along with the additive property of the group, it follows that $c(x)$ multiplied by any polynomial is also a codeword. For by iterating (9.36), $c(x)$ times any power of x is a codeword, and the sum of any such products is a codeword.

A cyclic code may be constructed by taking multiples of a generator polynomial $g(x)$:

$$c(x) = p(x)g(x) \tag{9.37}$$

If a binary code contains 2^k words, there must be an equal number of distinct polynomials $p(x)$, so the degree of $p(x)$ must be $k - 1$. The degree of $c(x)$ is $n - 1$, so the degree of $g(x)$ is $n - k$. It may be shown (Cattermole, 1986, p. 292) that there is a unique monic generator polynomial of minimum degree consistent with the required property, Eq. (9.37); that this does have the degree $n - k$; and that it is a divisor of $x^n - 1$.

A natural way to translate a k-bit data word into an n-bit codeword is to express the data word as a polynomial $p(x)$ and multiply by the generator, as in Eq. (9.37). We shall see later how this may be implemented.

Some of the codes discussed in Sec. 9.2 are cyclic codes:

1   The parity check code $(k + 1, k)$ has the generator:

$$g(x) = x + 1 \tag{9.38}$$

2   The Hamming (n,k) codes are cyclic codes whose generators are prime polynomials of degree $n - k$. Some examples of generators are:

$$(7,4) \text{ code: } g(x) = x^3 + x + 1 \tag{9.39}$$
$$(15,11) \text{ code: } g(x) = x^4 + x + 1 \tag{9.40}$$
$$(31,26) \text{ code: } g(x) = x^5 + x^2 + 1 \tag{9.41}$$
$$(63,57) \text{ code: } g(x) = x^6 + x + 1 \tag{9.42}$$

Prime polynomials for longer codes may be found in Cattermole (1986, p. 277) and in specialized texts.

3   Extended Hamming $(n + 1,k)$ codes have the basic Hamming generator multiplied by $1 + x$. Shortened Hamming codes have the generator appropriate to the next higher n, but the degree of $p(x)$ is reduced thereby dropping data bits. An example combining both features is a (7,3) code of distance 4, for which

$$g(x) = (x^3 + x + 1)(x + 1) \tag{9.43}$$

The codeword set may be exhibited easily by using the cyclic group properties along with the form of the generator. For example, the Hamming (7,4) code has the codeword set:

$$\begin{array}{ll} 0000000 & \\ 0001011 & \text{and its cyclic variants} \\ 1110100 & \text{and its cyclic variants} \\ 1111111 & \end{array} \tag{9.44}$$

Received vectors and error patterns may be represented by polynomials $r(x)$, $e(x)$ respectively, so that:

$$r(x) = c(x) + e(x) \tag{9.45}$$

An error is detectable if and only if $e(x)$ is not a multiple of the generator. The syndrome is the remainder on dividing the received word by the generator:

$$s(x) = R_{g(x)}[r(x)] \tag{9.46}$$

a polynomial of degree $n - k - 1$. This is characteristic of the error pattern, whatever codeword may have been sent. From Eqs. (9.45) and (9.46):

$$\begin{aligned} s(x) &= R_{g(x)}[c(x)] + R_{g(x)}[e(x)] \\ &= R_{g(x)}[e(x)] \end{aligned} \tag{9.47}$$

since the remainder left by $c(x)$ is zero.

Any detectable error has a nonzero syndrome. The syndrome (9.47) is nonzero for all errors within the weight bound (9.4). Also it is nonzero for all burst errors of length not more than $n - k$, since in this case $e(x)$

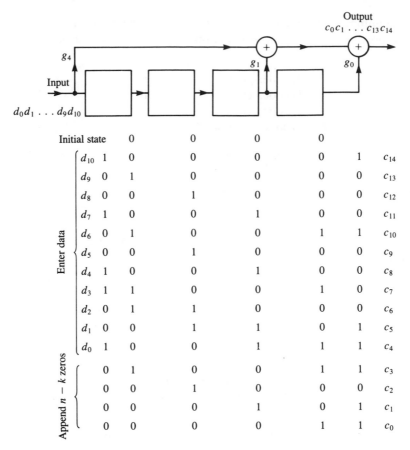

Initial state	0	0	0	0		
d_{10} 1	0	0	0	0	1	c_{14}
d_9 0	1	0	0	0	0	c_{13}
d_8 0	0	1	0	0	0	c_{12}
d_7 1	0	0	1	0	0	c_{11}
d_6 0	1	0	0	1	1	c_{10}
d_5 0	0	1	0	0	0	c_9
d_4 1	0	0	1	0	0	c_8
d_3 1	1	0	0	1	0	c_7
d_2 0	1	1	0	0	0	c_6
d_1 0	0	1	1	0	1	c_5
d_0 1	0	0	1	1	1	c_4
0	1	0	0	1	1	c_3
0	0	1	0	0	0	c_2
0	0	0	1	0	1	c_1
0	0	0	0	1	1	c_0

Figure 9.4   Cyclic encoder

is the product of a power of x with a polynomial of degree less than $n - k$, and this cannot be divisible by $g(x)$ which has degree $n - k$.

In an error-correcting code, all correctable errors must have distinct syndromes. It can be shown (Cattermole, 1986, p. 294) that the syndromes (9.47) are distinct for errors within the weight bound (9.55).

It is clear from Eqs. (9.37) and (9.47) that encoding and decoding correspond to polynomial multiplication and division, respectively. This suggests a mechanism. Multiplication of polynomials amounts to convolution of their coefficient sequences; in conventional signal processing this can be accomplished by a feedforward shift register, and the same is true for binary encoding provided that we use the appropriate field of coefficients, namely $GF(2)$. Figure 9.4 shows an example: the shift register connection is suitable for a Hamming (15,11) code, and the illustrative data entry shows the generation of:

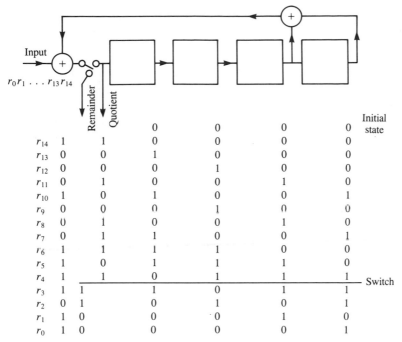

	Remainder	Quotient					
			0	0	0	0	Initial state
r_{14}	1	1	0	0	0	0	
r_{13}	0	0	1	0	0	0	
r_{12}	0	0	0	1	0	0	
r_{11}	0	1	0	0	1	0	
r_{10}	1	0	1	0	0	1	
r_{9}	0	0	0	1	0	0	
r_{8}	0	1	0	0	1	0	
r_{7}	0	1	1	0	0	1	
r_{6}	1	1	1	1	0	0	
r_{5}	1	0	1	1	1	0	
r_{4}	1	1	0	1	1	1	Switch
r_{3}	1	1	1	0	1	1	
r_{2}	0	1	0	1	0	1	
r_{1}	1	0	0	0	1	0	
r_{0}	1	0	0	0	0	1	

Figure 9.5   Cyclic decoder

$$(1 + x + x^4)(1 + x^3 + x^4 + x^7 + x^{10})$$
$$= 1 + x + x^3 + x^4 + x^5 + x^{10} + x^{14} \quad (9.48)$$

A syndrome polynomial is a remainder on division, so a decoder needs a division device. This can be implemented by a feedback shift register (Fig. 9.5). The divider can be understood as a multiplier into which trial values of a quotient $q(x)$ are fed: the trial values of $g(x)q(x)$ are subtracted from the input $r(x)$, and if division is exact the process continues until the remaining difference $r(x) - g(x)q(x)$ is zero. If division is not exact, the feedback loop is broken once $q(x)$ is complete, and the remainder is extracted from the open-loop structure. Thus the switch in Fig. 9.5 opens after the k-digit quotient has been fed into the shift register, and the next $(n - k - 1)$ bits to emerge are the remainder, namely the syndrome. The illustrative entry r in Fig. 9.5 is the codeword c of Fig. 9.4 save for one bit error. If there were no error, the remainder would be zero; the remainder 1100 in this example is characteristic of an error in digit 6.

The cyclic structure can be used to design powerful multiple-error-correcting codes, the problem being to find a generator $g(x)$ appropriate to the required properties. To understand these relationships, we need to develop a little more mathematical theory.

9.3.4 Spectral properties of fields and codes

We have several times referred to prime polynomials, meaning those with no factors having coefficients in the appropriate field. However, these can normally be factored if we admit elements from a larger field—the extension field. This is familiar from the relationship between real and complex numbers. For example, $x^2 + 1$ is prime if the coefficients are to be drawn from the real field, but composite if the coefficients may be drawn from the extension field of complex numbers:

$$x^2 + 1 = (x + j)(x - j) \qquad (9.49)$$

The use of an extension field is also familiar from Fourier analysis.[5] Let $[h_n]$ be a sequence of N real numbers. The transformed sequence:

$$H_k = \sum_{n=0}^{N-1} h_n w^{kn} \qquad (9.50)$$

where:

$$w = e^{-j2\pi/N} \qquad (9.51)$$

uses a kernel w in the complex field, and its terms may in general be complex. The real sequence may be arbitrary, but the complex sequence has Hermitian symmetry:

$$H_{N-k} = H_k^* \qquad (9.52)$$

The key property of the kernel w is that it is an Nth root of unity:

$$w^N - 1 = 0 \qquad (9.53)$$

We shall see that in a similar way, extension fields may be used in the analysis of finite-field constructs.

Firstly, we note that the possible extension fields of $GF(q)$ are $GF(q^m)$. The latter contain the elements of $GF(q)$, in the same way that a set of polynomials includes the coefficient field as a special case, namely the polynomials of zero degree. So our analysis of polynomials over $GF(2)$ will extend into $GF(4)$, $GF(8)$, $GF(16)$ and so on. Each of these fields will have a primitive element whose powers generate the remaining nonzero elements. If we denote a primitive element of the appropriate field by y, then:

$$
\begin{array}{lll}
x^2 + x + 1 & \text{is prime in } GF(2) & \\
& = (x - y)(x - y^2) & \text{in } GF(4) \qquad (9.54) \\
x^3 + x + 1 & \text{is prime in } GF(2) & \\
& = (x - y)(x - y^2)(x - y^4) & \text{in } GF(8) \quad (9.55)
\end{array}
$$

[5] It is assumed that the reader has some acquaintance with the discrete Fourier transform. A brief resume of relevant properties is given in Cattermole (1986, pp. 283–4). A more extended treatment is in Cattermole (1985, pp. 125–37; 1986, pp. 316–26).

$x^4 + x + 1 \qquad$ is prime in $GF(2)$
$$= (x - y)(x - y^2)(x - y^4)(x - y^8) \qquad \text{in } GF(16)$$
$$\tag{9.56}$$

These equations can be verified once we recall that the multiplicative group of the field is cyclic, and that successive powers of a primitive element y can be written as polynomials in x (see, for example, Cattermole, 1986, p. 280). The sequence of indices—all powers of 2—is no accident, as we shall see later. For a parallel, note that in Eq. (9.49) the roots $+j$ and $-j$ may be regarded as powers of a fourth root of unity.

Secondly, the familiar Fourier transform has a parallel in finite fields (Cattermole, 1986, p. 285ff). If $[h_n]$ is a sequence of N terms each taking values in a Galois field $GF(q)$, and y is an element of period N (i.e., an Nth root of unity) in an extension field $GF(q^m)$, then the Fourier transform of $[h_n]$ is a sequence $[H_k]$ in the extension field:

$$H_k = \sum_{n=0}^{N-1} h_n y^{kn} \tag{9.57}$$

which is formally identical with Eq. (9.50). The inverse transform needs to be specified with some care because of the difference in the fields (Cattermole, 1986, p. 285), but the parallel remains close, and all the familiar theorems of the classic Fourier transform have their counterparts in the Galois field, in most cases formally identical.

It follows that a real sequence, such as a codeword, has transform in an appropriate extension field, and that the concepts and methods of Fourier theory can be used to analyse encoding and decoding in much the same way as conventional Fourier theory is used to analyse conventional signal processing. In particular, we shall see that encoding and decoding of cyclic codes can be regarded as filtering operations.

The Fourier transform over the Galois field has conjugacy constraints relating subsets of components, analogous to the Hermitian symmetry of Eq. (9.52). They are (Cattermole, 1986, p. 287):

$$H_k^q = H_{kq} \qquad 0 \leqslant k \leqslant N - 1 \tag{9.58}$$

the index being interpreted modulo N. The implication is that components H_k may not be chosen independently: they are partitioned into conjugacy classes such that, if one member of a class be specified, the others are entailed. If we think of the H_k as frequencies, the classes may be called chords. The principle may be illustrated in the seven-point transform over $GF(2)$, which, since it requires a kernel of period 7, must use $GF(8)$, whose nonzero elements are a cyclic group of order 7. The chords have the following frequency content and constraint equations, derived from Eq. (9.58) with $q = 2$ and k ranging from 0 to 6:

$$\begin{aligned}
\text{Frequency } 0 \qquad & H_0^2 = H_0 \\
\text{Frequencies } 1, 2, 4 \quad & H_1^8 = H_2^4 = H_4^2 = H_1 \\
\text{Frequencies } 3, 6, 5 \quad & H_3^8 = H_6^4 + H_5^2 = H_3
\end{aligned} \qquad (9.59)$$

Note that the first chord can have only two values, the others eight values each, so the total number of vectors over $GF(8)$ which may be transforms is 128, the same as the number of binary vectors in the original domain.

One of the key properties of the Fourier transform is that multiplication in one domain transforms into convolution in the other. We have already remarked that multiplication of polynomials amounts to convolution of their coefficients. This suggests a relationship between transforms and polynomials which turns out to be very fruitful for coding theory.

Let the sequences in both domains be taken as polynomial coefficients, as in Eq. (9.32) *mutatis mutandis.* Then:

$$[F_k] = [G_k H_k] \qquad \text{implies} \qquad [f_n] = [g_n] * [h_n] \qquad (9.60)$$

whence:

$$f(x) = g(x)h(x) \qquad (9.61)$$

The basic definition of the transform, Eq. (9.57), may be put in polynomial notation:

$$H_k = h(y^k) \qquad (9.62)$$

Now the roots of the polynomial must be drawn from the extension field and so are powers of a primitive element y, and Eq. (9.62) implies:

$$\text{if } y^k \text{ is a root of } h(\ ), \text{ then } H_k = 0 \qquad (9.63)$$

This is consistent with (9.60) and (9.61), for if H_k is zero then F_k is zero, and the roots of the product $f(x)$ include those of $h(x)$.

Now let us apply these ideas to cyclic codes. By Eq. (9.37), any codeword $c(x)$ is a multiple of the generator $g(x)$ and so has the roots of $g(x)$. The foregoing theory implies that, if we express the generator in the transform domain as $[G_k]$, then all codewords $[C_k]$ have zero components wherever one appears in $[G_k]$.

From the conjugacy constraints, zeros come in chords; in Eq. (9.59), if H_1 is zero then so are H_2 and H_4. In the original domain, this implies that roots y, y^2 and y^4 should be associated, as indeed they are in Eq. (9.55). Prime polynomials are transforms of zero chords. Such polynomials, taken as generators, imply that the code has a zero chord in the transform domain.

These insights carry over into the encoding and decoding mechanisms for cyclic codes. The feedforward shift register of Fig. 9.4 has the same structure as a finite-impulse-response transversal filter. This encoder

h_0	h_1	h_2	h_3	h_4	h_5	h_6	H_0	H_1	H_2	H_3	H_4	H_5	H_6
0	0	0	0	0	0	0	0	0	0	0	0	0	0
1	1	1	1	1	1	1	1	0	0	0	0	0	0
1	1	0	1	0	0	0	1	0	0	0	α^4	α^2	α
0	1	1	0	1	0	0	1	0	0	0	1	1	1
0	0	1	1	0	1	0	1	0	0	0	α^3	α^5	α^6
0	0	0	1	1	0	1	1	0	0	0	α^6	α^3	α^5
1	0	0	0	1	1	0	1	0	0	0	α^2	α	α^4
0	1	0	0	0	1	1	1	0	0	0	α^5	α^6	α^3
1	0	1	0	0	0	1	1	0	0	0	α	α^4	α^2
0	0	1	0	1	1	1	0	0	0	0	α^4	α^2	α
1	0	0	1	0	1	1	0	0	0	0	1	1	1
1	1	0	0	1	0	1	0	0	0	0	α^3	α^5	α^6
1	1	1	0	0	1	0	0	0	0	0	α^6	α^3	α^5
0	1	1	1	0	0	1	0	0	0	0	α^2	α	α^4
1	0	1	1	1	0	0	0	0	0	0	α^5	α^6	α^3
0	1	0	1	1	1	0	0	0	0	0	α	α^4	α^2
1	0	0	0	0	0	0	1	1	1	1	1	1	1
0	1	0	0	0	0	0	1	α	α^2	α^4	α^3	α^5	α^6
0	0	1	0	0	0	0	1	α^2	α^4	α	α^6	α^3	α^5
0	0	0	1	0	0	0	1	α^3	α^6	α^5	α^2	α	α^4
0	0	0	0	1	0	0	1	α^4	α	α^2	α^5	α^6	α^3
0	0	0	0	0	1	0	1	α^5	α^3	α^6	α	α^4	α^2
0	0	0	0	0	0	1	1	α^6	α^5	α^3	α^4	α^2	α

Figure 9.6   Fourier transformation of Hamming (7,4) code

filters the data to impose spectral zeros on the coded version. The feedback shift register of Fig. 9.5 has the same structure as a complementary infinite-impulse-response transversal filter; it selects frequency components in the bandstop region of the encoder. Components in this region constitute a syndrome. No error implies zero syndrome; only if the signal has been modified in transmission do these components become nonzero. This is reminiscent of a well-known method of testing transmission systems for analogue distortions: send a broadband test signal with a slot imposed by a bandstop filter, then look for energy in the stop region generated by intermodulation.

Figure 9.6 tabulates the codewords of a Hamming (7,4) code together with their Fourier transforms (Cattermole, 1986, p. 304). Here α is a primitive element of $GF(8)$. The codewords have a zero chord (1, 2, 4), as would be expected from the generator (9.55). The one-bit error patterns, however, have distinct nonzero values in this region.

Multiple-error-correcting codes have more extensive sets of zeros, as we shall see in the following section.

9.3.5 Bose–Chaudhuri–Hochquenghem (BCH) codes

This is a class of multiple-error-correcting codes, invented by R. C. Bose and D. K. Ray Chaudhuri, and independently by A. Hochquenghem, some 10 years after the foundation of error control principles by R. W. Hamming and M. J. E. Golay. The BCH class includes as special cases the Hamming codes, and also the Reed–Solomon codes. Despite much further invention, this class remains of fundamental importance both in theory and in practice.

In terms of the theory developed in the previous section, there is a simple definition of a BCH code: it has a block of $2b$ consecutive spectral zeros, for some positive integer b. By careful manipulation of Fourier transforms, it may be shown that in the original domain the code has a distance $> 2b$, hence by (9.5) the potential for correcting b errors.

This last statement is so important in the theory of codes that it is worth trying to follow the reasoning. The following outline is based on a proof given in Cattermole (1986, pp. 305–6) which in turn is based on a proof by R. E. Blahut.

First we note that, by group symmetry, a distance greater than $2b$ follows if the weight of nonzero codewords is greater than $2b$. We then postulate a word of weight $w \leqslant 2b$ and show that it can only be zero, as follows. Let this word be $[c_n]$. Define a mask sequence $[m_n]$ such that $m_i = 0$ if and only if $c_i \neq 0$. Then on Fourier transformation we have:

$$[c_n][m_n] = 0 \quad \text{implies} \quad [C_k] * [M_k] = 0 \qquad (9.64)$$

Now by a principle like Eq. (9.63) with the domains interchanged, each zero component in $[m_n]$ corresponds to a root of the polynomial $M(x)$, so this has degree w, the weight of the postulated codeword. So the sequence $[M_k]$ takes the form $[M_0 \, M_1 \, M_2 \ldots M_w \, 0 \, 0 \ldots 0]$ with not more than $2b + 1$ nonzero components. This sequence enters into the convolution (9.64), which may be written in the recursive form:

$$M_0 C_k = - \sum_{j=1}^{2b} M_j C_{k-j} \qquad (9.65)$$

Now if $[C_k]$ has $2b$ consecutive zeros, Eq. (9.65) implies another zero, and then another, until $[C_k]$ is identically zero. Then $[c_n]$ is zero, i.e., the only word of weight $\leqslant 2b$ is the zero word, so the distance is greater than $2b$.

As examples, we consider 15-bit binary codes whose Fourier transform uses as kernel a primitive element of $GF(16)$. It may be deduced from Eq. (9.58) that the chords are:

$$
\begin{aligned}
&\text{A:} \quad \text{Frequency} \ \ 0 \\
&\text{B:} \quad \text{Frequencies 1, 2, 4, 8} \\
&\text{C:} \quad \text{Frequencies 3, 6, 12, 9} \qquad (9.66) \\
&\text{D:} \quad \text{Frequencies 5, 10} \\
&\text{E:} \quad \text{Frequencies 7, 14, 13, 11}
\end{aligned}
$$

Looking for a minimal set of chords to contain blocks of $2b$ frequencies, we have to take frequencies 1 to $2b$ consecutively and, incidentally, some

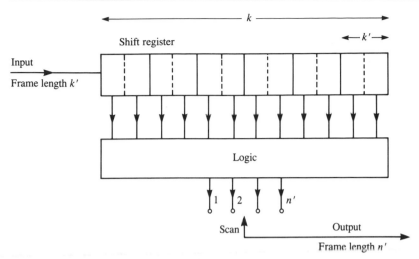

Figure 9.7 Encoder of rate k'/n'

other frequencies. The information content of the other chords, by Eq. (9.58), is one bit for every frequency included. Interesting 15-bit codes are:

$$
\begin{array}{lll}
(15, 11)\ \text{code:}\ b = 1 & \text{chord}\ B & = 0 \\
(15, 7)\ \ \text{code:}\ b = 2 & \text{chords}\ B,\ C & = 0 \qquad (9.67) \\
(15, 5)\ \ \text{code:}\ b = 3 & \text{chords}\ B,\ C,\ D = 0
\end{array}
$$

The first is of course the Hamming code; the others are BCH multiple-error-correcting.

BCH codes are widely used in practice. The Reed–Solomon codes are basically BCH codes of high radix: if the radix is a power of two, they may be mapped onto binary codes, and these also are widely used. Finally, we note that codes of these classes may be made more powerful still by concatenation.

9.4 Convolutional codes

9.4.1 The convolutional encoder

We noted in Sec. 9.1.3 that a possible procedure is to apply encoding operations to a running block of bits held in a shift register. A fairly general illustration of the principle is shown in Fig. 9.7. The incoming data bits are divided into frames each of k' bits (where k' is a small integer possibly unity). The data is entered, a frame at a time, into a shift register which stores k bits; k is a multiple of k' and is called the *constraint length*.[6] Following each shift, a codeword frame of n' bits is

---

[6] There are various definitions in the literature, all approximately though not all precisely equivalent to that given here.

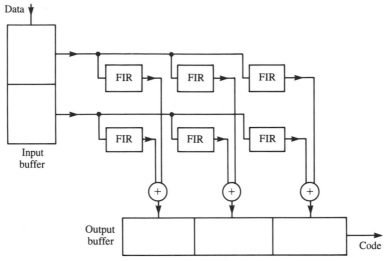

Figure 9.8   Equivalent structure for a convolutional encoder ($k' = 2$, $n' = 3$)

generated by logical operations on the shift-register contents. The rate of the code, defined as:

$$R = \frac{k'}{n'} = \frac{k}{n} \tag{9.68}$$

is a measure of the efficiency of the code, complementary to its redundancy. The rate is always less than unity, and typically a fraction in the range 1/4 to 3/4. The parameters k and n are length measures, similar to those denoted by the same symbols in our treatment of block codes in that n code bits are generated in response to every k data bits. However, the parallel is not precise. A long input sequence uniquely defines a long output sequence, but no subset of input bits uniquely defines the corresponding subset of output bits (with the obvious exception of a subset starting from quiescent initial conditions).

In a convolutional encoder, the logic is algebraically linear, that is, it can be realized entirely by modulo 2 adders. In consequence, the encoder can be considered as a set of $k'n'$ finite-impulse-response transversal filters. Each filter has for input the sequence of bits occurring at one of the k' positions of the input frame, and contributes via an adder to one of the n' bit positions of the output frame (Fig. 9.8). Now in our treatment of block codes, we noted that such a filter can be represented by a polynomial, and that passage through the filter multiplies an input polynomial (representing data) by a generator polynomial (representing the filter structure) to give an output polynomial (representing the code sent to line). Similarly, the structure of the convolutional code can be represented by a set of generator polynomials, one for each FIR filter. Algebraically, instead of the single equation:

$$c(x) = d(x)g(x) \qquad (9.69)$$

implemented by the structure of Fig. 9.4, we have:

$$c_j(x) = \sum_{i=1}^{k'} d_i(x)g_{ij}(x) \qquad (9.70)$$

where the subscript i identifies a bit position in the input frame, and the subscript j a bit position in the output frame. The input and output polynomials are now arbitrarily long: the generator polynomials are of finite degree, bounded by the constraint length.

Whereas Eq. (9.69) can be inverted to recover $d(x)$ from $c(x)$, using an infinite-impulse-response transversal filter as a divider (Figure 9.5), the superposition of several tributaries represented by the summation in Eq. (9.70) would appear to rule out a similar inversion. However, the data stream can be recovered from an error-free coded sequence by a slightly more complex inverse-filtering technique, if and only if the generator polynomials are relatively prime; the mathematical basis for this is the Chinese remainder theorem (Cattermole, 1986, p. 327). A further reason for using relatively prime polynomials is that a damaging form of error propagation may be avoided (Blahut, 1983, p. 360).

9.4.2 Code structure and distance

The geometric view of block codes and their distances given in Sec. 9.2.1 does not apply directly to convolutional codes: the effective 'block' of a convolutional code is continually shifting. We need a pictorial form in which sequences can be identified with paths, in an open-ended manner. A useful form is the trellis diagram, of which an example is shown in Fig. 9.9. For simplicity, we illustrate the special case $k' = 1$, which in practice is of common occurrence. An encoder of constraint length k will then have a single shift register of k stages. As each new bit is entered, $k - 1$ bits are shifted and the oldest bit is lost. We distinguish 2^{k-1} states, identified by the $k - 1$ bits which are retained and entered in the trellis diagram at distinct vertical levels. The columns, moving from left to right, relate to successive bit times. Edges identify state transitions which may occur as new bits arrive. For example, let $k = 4$ and suppose the initial register contents to be $101x$. On arrival of a 0 bit, the contents change to 0101 whereas on arrival of a 1 bit they change to 1101. So from state 101 there are two possible transitions, to 010 and 110 respectively. With the numbering used in the diagram (binary number sequence, with the latest bit as least significant) the two alternative successor states are at adjacent levels, with a 0 bit arrival

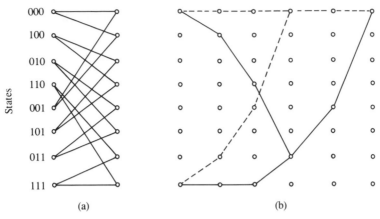

Figure 9.9   Trellis diagram

selecting the upper and a 1 bit arrival the lower. Figure 9.9(a) shows the transitions for $k = 4$: the generalization should be obvious.

A sequence of input bits causes a sequence of transitions represented by a path on the trellis diagram. For example, Fig. 9.9(b) shows the paths corresponding to inputs 00000 (broken line) and 11000 (solid line) starting from states 000 and 111.

After each entry of an input bit, an output frame of n' bits is generated by linear algebraic operations on the new contents of the shift register. Each transition in the trellis diagram uniquely defines a frame, and can be labelled accordingly. To illustrate this, we take the specific $(n,k) = (8,4)$ encoder shown in Fig. 9.10(a) with generator polynomials:

$$g_1(x) = 1 + x + x^2 + x^3$$
$$g_2(x) = 1 + x + \quad\ \ x^3 \tag{9.71}$$

It is easily deduced from the logic that the two-bit frames corresponding to the transitions are those indicated in Fig. 9.10(b). The coded output can be constructed for any given data input and given initial state, for example those of Fig. 9.9(b), as shown in Table 9.1.

This example illustrates two important points. Firstly, the code output generated in response to a specific input is not unique but depends on

Table 9.1 Coded output for the data inputs and initial states of Fig. 9.9(b)

Data	Code with initial state 000					Code with initial state 111				
00000	00	00	00	00	00	10	01	11	00	00
11000	11	00	01	01	11	01	01	10	01	11
Difference of codes	11	00	01	01	11	11	00	01	01	11

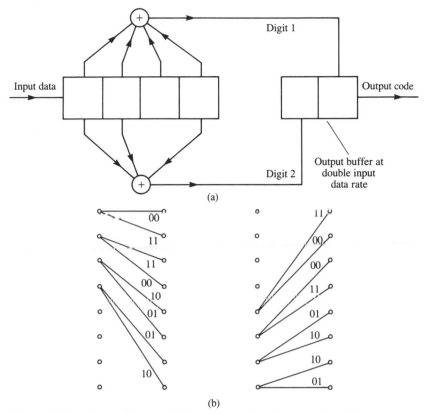

Figure 9.10 An (8,4) convolutional encoder: (a) encoder, (b) codes associated with transitions

past history, insofar as this affects the initial state. Secondly, the difference between codes generated by different data inputs is uniquely defined: this follows from the algebraic linearity of the system.

In block codes, the error-correcting power depends in a simple way on the distance between codewords (expressions 9.4–9.6), and the critical distance of a code is the minimum distance between any two codewords. Moreover, because of group symmetries, we need investigate only the distances from the zero codeword: the minimum distance equals the minimum weight of a nonzero codeword. Similarly, in a convolutional code we can define distance in terms of the deviation from an all-zero path. Consider a path segment which diverges from the zero path and later returns to it, such as the upper solid-line path in Fig. 9.9(b). The distance of that path segment from the zero path is the weight of the code generated, in this case 6 (the codeword is in the table above). The *free distance d* of a convolutional code is the minimum weight attached to any such path segment. In fact, the free distance of the code in this example is 6, and the segment shown is of minimal weight. This property is easily verified by exhaustive search. There is also an analytic method for deriving a generating function for the distribution of distances, and hence finding the free distance (Sklar, 1988, pp. 339–41).

A code with free distance d offers the possibility of correcting up to b errors within the span of a path segment of the type described, if:

$$2b < d \tag{9.72}$$

It is not possible to state the error-correcting power as precisely as for block codes, because: (1) the critical segment length is not simply related to the code parameters; (2) errors may propagate as a result of previous history; and (3) practical decoders are neither perfectly nor equally efficient. Perhaps the strongest general statement is that a good decoder will, with high probability, correct $b < d/2$ errors occurring within a few constraint lengths.

A good code therefore has a free distance which is as large as possible for a given rate and constraint length. Unfortunately no general method is known for constructing such codes. A class of distance 3 codes has been defined in terms of a general theory (Wyner–Ash codes), analogous and indeed closely related to the Hamming codes. However, its usefulness is very limited: as we noted in Sec. 9.3.1, distance 3 can be useful in ARQ systems but is quite inadequate for FEC systems, and convolutional codes are, by their nature, restricted to the FEC mode. Most codes with practically useful parameters have been found by computer search. For example, Blahut (1983, p. 367) tabulates optimal codes for constraint lengths 3 to 14 and rates 1/2, 1/3 and 1/4. The best attainable free distance d increases irregularly with k. It may be approximated within one unit by the empirical expressions:

$$
\begin{align}
R = 1/2 \qquad & d \approx \quad k + 2 \tag{9.73} \\
R = 1/3 \qquad & d \approx 1.6(k + 2) \tag{9.74} \\
R = 1/4 \qquad & d \approx 2.2(k + 2) \tag{9.75}
\end{align}
$$

Note that our example above ($n = 8$, $k = 4$, $d = 6$) satisfies (9.73) as an equality.

9.4.3 The convolutional decoder

Before describing a practical decoding procedure, we speculate for a moment on an ideal decoder which represents an upper bound to the performance attainable. We noted in Sec. 9.4.1 that for a large class of codes, formal inversion of the encoding equations (9.70) is possible, but (unlike the division process in block codes, which terminates) the process continues indefinitely. An optimum decoder would operate on the entire received sequence, from the beginning, and make a maximum-likelihood decision about the message sent. If we assume that each received digit is detected independently, the decoder would seek a possible transmitted sequence with minimum Hamming distance from the received sequence, effectively treating the whole transmitted sequence as one very large block.

Such an ideal procedure is not feasible. The convolutional decoder, like the encoder, operates on a running block of digits. It has a 'decoding window' of bounded length, which advances along the incoming bit stream. How long must this decoding window be, for reliable decoding? And what decoding procedure is to be used? There are several known procedures, the choice permitting a trade-off between reliability of performance and decoding complexity. We describe only one procedure, which is known to be asymptotically maximum-likelihood as the decoding window is extended.

It is immediately clear that the decoding window must be longer than the constraint length, for the span of a segment illustrating the free distance is normally greater, as in our example of Figs. 9.9(b) and 9.10. Furthermore the decoder—unlike the encoder—does not know the initial state in relation to which a coded sequence is to be interpreted, and to establish the current state a full constraint length of correct decoding is required. We shall give an example in which the required window is three to four constraint lengths; about five constraint lengths is common in practice.

A true maximum-likelihood decoder would compare the code associated with each possible path through the trellis with the received signal, calculate the Hamming distance in each case, and select the path at minimum distance from the signal. A good practical decoding procedure should approximate the same result while operating on a running basis, with a minimal number of comparisons. The *Viterbi algorithm* reduces the comparisons, according to the following principles.

At each step through the trellis, previously separate paths will merge (as for example in Fig. 9.9b). A pair of merging paths may exhibit different Hamming distances from the received signal. The path with greater distance may be eliminated from further consideration: it cannot form part of a minimum-distance sequence. The path with lesser distance is retained and enters into later comparisons: this path is called the survivor. If two merging paths have the same distance, then either may be chosen as the survivor and the other discarded. There is then just one surviving path leading to each state of the system, after any step. On the next step, just two paths merge into any state, and again a comparison must be made and a survivor chosen. The procedure continues indefinitely, step by step.

The procedure is most readily understood by working through an example. Figure 9.11 gives decoder trellis diagrams for several cases, all based on the encoder of Fig. 9.10. Figure 9.11(a) shows the survivors after five steps if the all-zero code is sent and correctly received. The

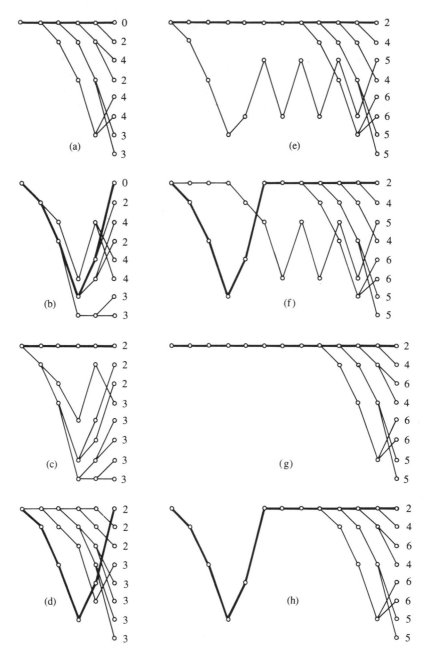

Figure 9.11 Decoding—surviving paths in a Viterbi algorithm example: (a) encoded 00000, no errors; (b) encoded 11000, no errors; (c,e,g) encoded 00000 . . ., initial double error; (d,f,h) encoded 11000 . . ., initial double error

paths form a tree, starting from the 000 state and branching out to all
states. The numbers are the distances pertaining to each path, that is,
the Hamming distance between the code associated with the path, and
the (all-zero) code received. The true path associated with the coded
signal is distinguished in the diagram by bold lines. Of course, the true
path could not at this stage be identified using only information available
to the decoder. However, the first arc of the true path does have a
distinctive property, identifiable at the decoder: it is common to all eight
surviving paths. So this first arc is common to all possible minimum-
distance solutions, and the first digit can be decoded, as a zero. It is easy
to verify that at the next step a valid tree of survivors comprises a
similar pattern of branches shifted one place to the right, with the
common stem now comprising two arcs, so that a further digit could be
decoded: the process could continue indefinitely.

Figure 9.11(b) shows a similar decoding tree for correct reception of a
coded sequence for the data 11000. Again there is a common stem,
identifying the first digit as 1. Continuation of the process yields as a
common stem the correct path (which is distinguished in the figure by
bold lines). Note that the decoding trees in Fig. 9.11(a) and (b) are
isomorphic, on which we shall comment later.

The two valid paths of these examples are those which we previously
used in Fig. 9.9 to illustrate the code's free distance of 6. So, by (9.72), it
should be possible to correct a double error within the span of these
paths. Figure 9.11(c) and (d) show the trees of surviving paths on the
assumption that the first two received bits are in error (the received
sequences from Table 9.1 then being 1100000000 . . . and 0000010111
. . . respectively). We see that after five steps the trees have no common
stem, and so no digits can be decoded reliably. Indeed, the uncertainty
remains after as many as 11 steps, whose survivors are shown in Fig.
9.11(e) and (f); most of the paths have a common stem, but one remains
obstinately distinct. On the 12th step, however, the odd path disappears,
and a common stem comprising eight arcs becomes apparent (Fig.
9.11(g) and (h)). As a further example (not illustrated) if either one of
the first two bits is in error, uncertainty prevails up to the seventh step
but on the eighth step a common stem of four arcs emerges.

How far do these examples demonstrate a general property? We have
remarked that the decoding trees of Fig. 9.11(a) and (b) are isomorphic;
further isomorphic pairs are (c) and (d), (e) and (f), (g) and (h). These
relationships, like some of those in the previous section, follow directly
from the group symmetries induced by the linearity of the algebraic
operations. Several conclusions follow:

1   With a given error pattern, any coded sequence would give rise to an
    isomorphic decoding tree characteristic of the error pattern.

2   Our examples start from the state 000, but any initial state would

give isomorphic decoding trees provided that previous decodings had been correct.

3 Once a common stem has emerged, and correct decoding commenced, the decoding tree follows a standard form (compare the branching portions of Fig. 9.11(a), (g) and (h)): consequently, the decoder handles any further errors independently of those which have just been corrected.

It would seem, therefore, that our small set of examples is properly representative of the behaviour of a decoder faced with error bursts which individually are within its correcting power (as judged by free distance) and are sufficiently well separated.

We can now see why the Viterbi algorithm is asymptotically maximum-likelihood: if carried back far enough, it utilizes all the information necessary to find a minimum-distance path. However, as these examples show, the delay in decoding varies with the error pattern: if the process is curtailed at any predefined depth, there is a possibility of decoding failure. A practical procedure is to store surviving paths to a fixed depth, say four or five constraint lengths, and to take as output at each step the bit corresponding to the unique first arc if, as in our examples, the paths have a common stem, or otherwise the first arc on the most probable path or set of paths. For easy understanding we have exhibited the process graphically, but of course in practice the paths are stored as a data array.

Further aspects of convolutional encoding and decoding are described in Blahut (1983) and Sklar (1988); the Viterbi algorithm is formally analysed in Chapter 5. These references also cite primary sources.

9.5 Applications of error control coding

Error control coding is widely but by no means universally used in digital systems. It is quite unnecessary for digital speech conveyed over the mainstream digital network: the error rate of the medium is very low, and the message has so much internal redundancy that an occasional error has little subjective effect. Error control is useful if either of these favourable conditions is breached: if the value of the message is critically dependent on reliable transmission, or if the transmission medium is notably prone to error.

Messages requiring high reliability include:

- Digital data with no internal redundancy,
- Control signals for multiple-user systems of all kinds,
- Voice or video signals which have undergone highly efficient source coding.

Transmission media with potentially high error rates include:

- The switched telephone network used for the conveyance of voice-frequency data signals,
- Very long radio paths, in space communication,
- Radio paths particularly vulnerable to fading, as in mobile systems,
- Storage media which may exhibit localized flaws, leading to error bursts in recovered data.

The first and still the predominant use of error detection and ARQ is for computer communications; transactions in which the terminals are normally utilizing stored data, retransmission is feasible, and delay is not too critical. Such systems use block codes as described in Secs. 9.2 and 9.3.

The first use of FEC was in space communications, where long radio paths yielded very poor signal-to-noise ratios and long propagation delays prevented the use of ARQ. Multiple-error-correcting codes were used, notably orthogonal and biorthogonal codes (highly redundant codes not described here) and codes of the BCH family described in Sec. 9.3. Historically, this application provided a stimulus for much interesting development of algebraic coding theory.

It has long been usual to protect system control signals by high redundancy, sometimes by simple repetition, now increasingly by more complex coding.

Some of the more recent applications of error control have introduced two factors which were less prominent in early work: the occurrence of quite long burst errors (typical both of fading media and of storage media with dropouts); and the need to process signals on a running basis at high speeds. The latter in particular encourages the use of convolutional codes, which are described in Sec. 9.4. Burst errors require the protective code to span a long period; while some simple codes (both block and convolutional) have been designed for this purpose, a very important technique is that of concatenation and interleaving. A simple example of this is given in Sec. 9.2.3. Much more powerful codes may be concatenated with advantage, to gain the effect of very long blocks.

A good current example of concatenation is the compact disc audio system, which employs cross-interleaved Reed–Solomon codes. We recall that the RS code is basically similar to the BCH code, but with high-radix symbols; in this case, each symbol is an eight-bit byte. The CD system concatenates (32,28) and (28,24) RS codes, effectively (256,224) and (224, 192) in binary terms, interleaved to produce an effective block length of 57 344 bits. Bursts as long as some 4000 bits can be corrected, and even longer ones detected for the purpose of masking errors by interpolation between the nearest reliable samples, exemplifying remarkably effective code design and implementation.

Another current example of complex coding is the GSM digital cellular mobile telephone system. In a real-time channel, ARQ is not feasible, and moreover the processing delays of very long block codes would be a disadvantage, but with a medium prone to fading, burst errors must be corrected. The GSM system uses a combination of short blocks, elaborate interleaving and convolutional coding for speech channels, together with, for data links over speech channels, a superimposed ARQ protocol with constraints on retransmission delay.

Many more examples could be found among transmission, storage, and processing systems. But perhaps enough has been said to suggest that coding is no longer just an esoteric speciality; some understanding of its principles and possibilities should be widespread among communication engineers.

References

Blahut, R. E. (1983) *Theory and Practice of Error Control Codes*, Addison-Wesley.

Cattermole, K. W. (1985) *Mathematical Foundations for Communication Engineering*, vol. 1, Pentech Press, London.

Cattermole, K. W. (1986) *Mathematical Foundations for Communication Engineering*, vol. 2, Pentech Press, London.

Sklar, B. (1988) *Digital Communications*, Prentice-Hall.

10 Line codes

EDWIN JONES

10.1 Definition

A *line code* defines the equivalence between sets of digits generated in a terminal and the corresponding sequence of symbol elements transmitted over a channel. It must be chosen to suit the characteristics of the particular channel. For metallic cables which have an essentially low-pass frequency characteristic, it is usually the last coding function performed before transmission. For band-pass channels, such as optical fibres, radio systems, and the analogue telephone network, where modulation is required before transmission, the line coding function either immediately precedes the modulator or is incorporated within the modulation process. The relationship between line coding and other channel coding functions is shown in Fig. 10.1. From this we note that the line coder and the corresponding decoder at the receiving terminal may have to operate at the transmitted symbol rate, in which case a reasonably simple design will usually be essential, especially for high-speed systems.

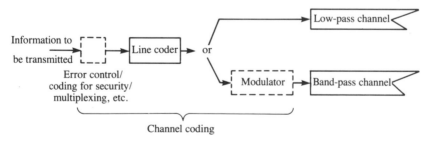

Figure 10.1 Relationship between line coding and other channel coding functions

10.2 Purpose

A line code provides the transmitted symbol sequence with the necessary properties to ensure reliable transmission and subsequent detection at the receiving terminal. To achieve this the following have to be considered:

- *Spectrum at low frequency* Most transmission systems are easier to design if a.c. coupling is used between waveform processing stages. This means that decoding cannot usually rely on receiving a d.c. spectral component. Furthermore, the coupling circuit components (capacitors or transformers) introduce a low frequency spectral cut-off and this puts a limit on the permissible low-frequency content of the line code if long-term intersymbol interference is to be avoided (Cattermole, 1983).
- *Transmission bandwidth required* If transmission bandwidth is at a premium, a multilevel line code, in which each transmitted symbol represents more than one bit of information, will enable the transmitted symbol rate to be reduced. However, for a given bit error ratio, a multilevel line code will require a better signal-to-noise ratio than a corresponding binary code.
- *Timing content* There must always be sufficient embedded timing information in the transmitted symbol sequence so that the distant receiver (and intermediate repeaters if used) can extract a reliable clock to time their decision-making processes. This usually means ensuring that the line code provides an adequate density of transitions in the transmitted sequence.
- *Error monitoring* By adding redundancy into the information stream a line code can provide a means of in-service monitoring of the error rate of its transmission link. For example, a line coder can be constrained so that it never produces certain symbol sequences; the occurrence of these illegal sequences at the receiver will then provide a means of estimating the link error performance.
- *Efficiency* To provide the above features it is usually necessary to add extra information (redundancy as far as the data is concerned) into the transmitted digit stream. This lowers the efficiency of the line code which is defined as:

$$\text{Efficiency} = \frac{\text{average information carried per transmitted symbol}}{\text{maximum possible information per symbol}} \times 100\%$$
$$\text{(when assuming no added redundancy)}$$

10.3 Classification of line codes

Line codes can be classified in a variety of ways, one method is to identify the following categories:

- Bit-by-bit codes
- Block codes
  —Bit insertion
  —Block substitution

- Partial response codes

Some codes fall into more than one of these classes. Many line codes exist, each designed to meet the needs of particular transmission systems. To illustrate the variety of design strategies available, we now consider representative tutorial examples from each category.

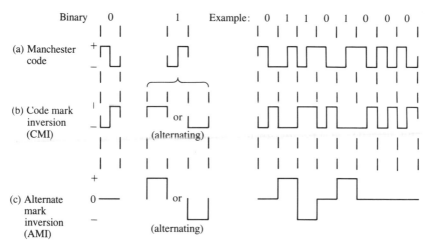

Figure 10.2   Some bit-by-bit line codes

Manchester code

This code (also called biphase or WAL1) is a simple example of a bit-by-bit code. Each information bit is coded into a two-bit symbol for transmission, as shown in Fig. 10.2(a). No d.c. component is generated because the polarity and amplitude of the first half of each transmitted symbol is complemented by the second half. Also, the resulting signal transition in the centre of each symbol ensures an adequate clock content. The absence of such a transition indicates a transmission error and so error monitoring is also possible. However, only one bit of information is carried for every two bits transmitted and so the efficiency is only 50 per cent. In this case, this results in a channel bandwidth requirement of twice the uncoded bandwidth. However, its simplicity makes it attractive for applications where transmission bandwidth considerations are less crucial; it is used in the magnetic recording of digital signals and in the Ethernet local area network system (as detailed in the ANSI/IEEE standard 802.3).

Code mark inversion (CMI)

As defined in Fig. 10.2(b), this code is similar to the Manchester code, except that it achieves a polarity balance over a longer term by alternating between a positive symbol and a negative symbol when a binary 1 (mark) is to be transmitted. It too is an agreed standard (CCITT, 1988) where it is specified for interface purposes when connecting together certain types of transmission equipment over short distances where cable bandwidth is unlikely to be a limiting factor.

Alternate mark inversion (AMI)

This is also a bit-by-bit code in that it codes one bit of data at a time, but it differs from the above two codes in that the transmitted symbol rate is the same as the binary information rate. It provides the desired features by using a redundant signal set with three possible amplitude levels, as shown in Fig. 10.2(c). The zero d.c. property is achieved by sending marks as a positive or negative symbol alternately. An estimate of the link error performance is obtained by counting violations of this alternating mark inversion rule. We note that AMI has a potential weakness; if many successive zero-level symbols are transmitted, then the clock will fail. Thus, either the number of successive zeros has to be restricted, which is undesirable for a transparent data link, or a code substitution has to be made to prevent this potential hazard. Such AMI-based substitution codes are used in 24- and 30-channel PCM telephone systems, where they are known as B6ZS and HDB3, respectively (see, for example, Waters, 1983).

mB1C

This code (Yoshikai et al., 1984) is an example of a block code of the bit insertion type. A block of m information bits has added to it a single extra bit which is selected in an attempt to provide a long-term polarity balance (i.e., an equal number of positive and negative transmitted pulses) and so no d.c. component. For this to work reliably, the source data may need to be scrambled first to ensure that it is adequately randomized. The added bit also ensures a minimum clock content and some redundancy to permit transmission error monitoring. Binary codes of this type feature high efficiency (m has ranged in practice from 7 to at least 23) and simple hardware, and are thus attractive for high-speed applications such as optical fibre systems. Codes such as mB1P (Dawson and Kitchen, 1986) and DmB1M (Kawanishi et al., 1988) are similar in concept but offer slightly different features.

mBnB

These codes (Brooks and Jessop, 1983) are block codes of the block substitution type where m binary source bits are mapped into n binary

Input (7 bits)	Output as transmitted (eight bits)			
	Negative disparity	Zero disparity	Positive	← word disparity
1111111		$- + - + - + - +$		
.				
.				
.				
1010001	$+ - - - + - - +$	or	$- + + + - + + -$	
.				
.				
.				
0101001	$- + - - - + - -$	or	$+ - + + + - + +$	
.				
.				
0000000		$+ + + - + - - -$		

Figure 10.3   Part of the translation table for 7B8B

bits for transmission. Redundancy is built into the code to provide the desired transmission features by making $n > m$. Several such codes have been proposed (and used), in particular where $n = m + 1$. They differ from the previous category in that the n-bit transmitted block may bear little similarity to its input source block. This gives greater flexibility for providing the desired line code features for a given level of code efficiency, but it is achieved at the expense of increased encoder and decoder circuit complexity.

By way of example, part of the translation table for the balanced polarity 7B8B code (Sharland and Stevenson, 1983) is illustrated in Fig. 10.3 where it can be seen that each seven-bit input source word is mapped into one or two possible eight-bit output words. Output words that are balanced in themselves are said to have zero *disparity* (or inbalance) and are to be preferred in that a single input-to-output mapping is sufficient (as seen in the top and bottom entries of the table in Fig. 10.3). However, there will only be a limited number of these balanced output words and so the table has to be completed by using nonzero disparity output word pairs where members of each pair have a disparity of opposite sign. Only one of these words is selected for transmission; the selection is made on the basis of minimizing the *cumulative disparity* over the transmitted sequence. It follows that some possible output words will not be needed, and this redundancy provides the necessary design flexibility. The selection of words and their mappings from input to output is made on the basis of ensuring good timing content, error monitoring, and word alignment, and of minimizing the opportunity for transmission error multiplication in the decoding progress. A computer search may be used to optimize this mapping. Many such binary-to-binary balanced codes have been designed.

4B3T

This code (Catchpole, 1975) maps four binary input bits into three ternary output symbols as shown in the translation table of Fig. 10.4. (The ternary output symbols are labelled as $+$, 0 and $-$). It can be seen that it uses a similar design philosophy to the 7B8B code, in that if the output word is not balanced in itself, then a complementary version is always available. A disparity control counter then selects for transmission whichever version is required to ensure a long-term polarity balance (i.e., no d.c. signal component) on the transmission line. Unlike AMI, 4B3T uses the extra information capacity available with multilevel transmission to reduce the line symbol rate (to 3/4 of the input binary rate in this case). However, there is still a little redundancy in that, as can be seen from Fig. 10.4, only 26 out of 27 (3×3^2) possible three-symbol ternary output words are used. This means that the word 000 need not be assigned, which ensures that whatever the input data sequence, there will always be an adequate density of symbol transitions in the line signal for timing purposes. This redundancy, although small, can also be used, together with information gained from apparent violations of the disparity controlling rules, to monitor transmission errors and to spot word misalignment problems at a decoding terminal.

   In an attempt to optimize the properties of the code, a variety of input-to-output mappings have been published, Figure 10.4 is an early one. Other variants are also known, for example MS43 (Franaszek, 1968) provides a more sophisticated choice of output codewords. This permits a better short-term code balance, which in turn reduces the low-frequency spectral content.

Input (four bits)	Output as transmitted (three ternary symbols)			Output word disparity
	Negative	Zero	Positive	
1111		$- 0 +$		0
1110		$0 + -$		0
1101	$0\ 0\ -$		$0\ 0\ +$	± 1
1100	$0 - 0$		$0 + 0$	± 1
1011	$- 0\ 0$		$+ 0\ 0$	± 1
1010	$- + -$		$+ - +$	± 1
1001	$+ - -$		$- + +$	± 1
1000	$- - +$		$+ + -$	± 1
0111	$- - -$		$+ + +$	± 3
0110	$- 0 -$		$+ 0 +$	± 2
0101	$0 - -$		$0 + +$	± 2
0100	$- - 0$		$+ + 0$	± 2
0011		$+ - 0$		0
0010		$0 - +$		0
0001		$- + 0$		0
0000		$+ 0 -$		0

Figure 10.4   A translation table for 4B3T

2B1Q

This is a block code that moves away from the redundancy-based design methods discussed so far. It converts two binary input bits into a single four-level (quaternary) symbol, as shown in Fig. 10.5. It follows that, for a random data input, there will be no unused transmission sequences and so none of the redundancy by which the examples so far have provided the requisite line coding features. Thus, to ensure that the input data is unlikely to result in long sequences of similar transmitted symbols (and so cause a loss of d.c. balance and timing problems) the data has to be adequately scrambled before encoding. The code therefore features 100 per cent efficiency, simple encoding and decoding (the latter without needing word synchronization), and (with good scrambling) reliable d.c. balance and timing content. However, with no redundancy, it does not permit link error monitoring; if required, this has to be provided by other means. Relative to the uncoded binary input, 2B1Q permits a 50 per cent reduction in symbol rate and so is attractive for bandwidth and crosstalk limited transmission links. For these reasons it has become an interface standard (ANSI, 1988) for subscriber line access to an integrated services digital network (ISDN).

Input (two bits)	Output as transmitted (one quaternary symbol)
10	+3
11	+1
01	−1
00	−3

Figure 10.5 A translation table for 2B1Q

Partial response codes

These codes (Lender, 1981), known as *correlative* or *multiple response codes*, deliberately introduce a controlled amount of i.s.i. into the received signal. This known amount of correlated interference produces a multilevel received signal which, with suitable pretransmit coding, can be designed to provide the requisite line code features.

A simple example of the class is duobinary (Lender, 1966), in which a binary signal can be transmitted at twice the rate required to give zero i.s.i. With suitable coding and equalization the resulting i.s.i. can be arranged to give a three-level received signal. Figure 10.6 gives illustrative waveforms for this code. Figure 10.6(a) shows an input step applied to a transmission channel together with the output required at the receiver after equalization. It is seen that the output has a relative amplitude of 0.5 after time $T/2$ and 1.0 after time T (ignoring channel delay). Now, if

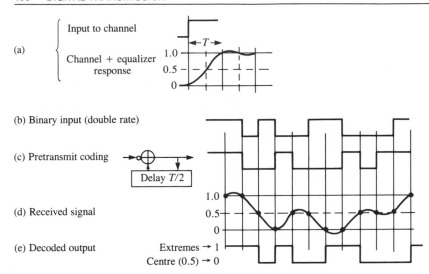

Figure 10.6   Duobinary—a partial response code

signals are launched into the channel with a time spacing of less than T, the output will have time to respond only partially to the input stimuli. Figure 10.6(b) shows such an input where the symbol period is only $T/2$, i.e., signalling is at twice the usual rate permitted for zero i.s.i. With duobinary, to permit bit-by-bit decoding and so prevent error multiplication at the receiver, some pretransmit coding is used as detailed in Fig. 10.6(c). This coded signal is transmitted over the channel and through its associated equalizer. The result is shown in Fig. 10.6(d). The received signal can then be applied to a three-level digital decision circuit and decoded, bit by bit, in accordance with the rules given in Fig. 10.6(e) to yield an output which is the same as the original binary input. Thus, a binary data signal has been transmitted at twice the conventional rate, the price paid being careful equalization and a three-level signal at the receiver. The code features adequate transitions for clock extraction and also some redundancy suitable for transmission error monitoring (by noting in particular that the output cannot change between the two extreme levels in adjacent decision times). Unfortunately, duobinary is not a d.c.-free code because the two extreme levels cannot be guaranteed to balance each other. However, *modified duobinary* (Lender, 1966) can provide the requisite balance property but at the expense of a narrower received eye because, unlike duobinary, direct transitions between the two extreme received signal levels become possible.

   For well-behaved channels (where i.s.i. can be accurately controlled), partial response codes can provide a bandwidth-efficient transmission technique. They have been used in radio systems and also to convert existing 24-channel PCM systems into 48-channel systems with a minimum of equipment change.

10.4 Choosing a line code

The above examples demonstrate a variety of ways of ensuring that a line code has the necessary properties. A fundamental factor to consider when choosing a code for a particular application is the opportunity, in principle at least, to trade transmission bandwidth against the number of amplitude levels in the transmitted symbol. By using a multilevel code we can increase the information carried per symbol and so reduce the transmitted symbol rate and hence reduce the required transmission bandwidth. We can derive a bound on this trade-off by noting that an m-level symbol can carry a maximum of $\log_2 m$ bits of information per symbol. Thus, an m-level code would permit, in theory, a corresponding reduction in symbol rate and so a similar reduction in channel noise entering the receiver. However, as seen in Chapter 1, an m-level received signal needs an $(m^2 - 1)/3$ improvement in signal-to-noise ratio relative to binary for the same error probability. Thus we have a bound for m-level transmission:

Signal-to-noise ratio penalty (relative to binary)

$$= \frac{\text{received signal-to-noise ratio improvement factor required}}{\text{noise bandwidth reduction factor}}$$

$$= \frac{m^2 - 1}{3 \log_2 m}$$

This is plotted (in decibels) in Fig. 10.7 which shows that multilevel transmission incurs an increasing signal-to-noise ratio penalty as m increases, i.e., binary is best. However, this bound is based on the receiver noise bandwidth assumption made above, but other factors may also have to be considered. For example, channel bandwidth may be insufficient to support binary transmission and so multilevel coding must be used. Similarly, crosstalk in multipair metallic cables rises rapidly with transmission rate, when this fact is linked with transmission bandwidth limitations a lower symbol rate code may perform better, these are the conditions which favour the 2B1Q code outlined earlier.

Equalization is another factor which often needs to be considered when choosing a line code. The partial response codes mentioned earlier rely on controlled amounts of i.s.i. to provide the requisite received waveform: is the precoder in duobinary (Fig. 10.6c) a line coding function or an equalizer? Other codes also blur this distinction, for example the Manchester code mentioned earlier can provide a worthwhile amount of self-equalization as can be seen in Fig. 10.8. Figure 10.8(a) is derived from Fig. 1.6 and shows the result of transmitting individual 10 µs pulses over metallic pair cables of different lengths. (To aid comparison, the amplitudes of the received waveforms have been scaled and their transmission delays adjusted so that all signal

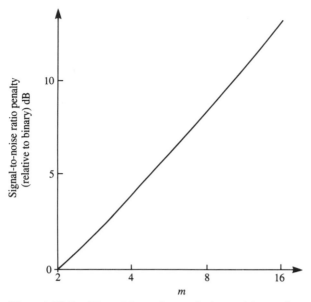

Figure 10.7   Signal-to-noise ratio bound for *m*-level transmission

peaks are coincident). We note the prohibitive amount of i.s.i. that
would be incurred if we attempted to transmit 10 μs spaced pulses
without equalization. However, if each pulse is replaced with a
Manchester-coded (positive then negative) symbol, then as Fig. 10.8(b)
shows, for the longer cable lengths a useful amount of *tail cancellation*
occurs so that i.s.i. is considerably reduced. This partial self-equalizing
property means that for this code little or even no equalization may be
needed. Some other line codes can act in a similar fashion.

   In summary, it can be seen that choosing the right line code for a
particular application is not always easy. Many interrelated factors have
to be considered, with the objective of providing a code with satisfactory
spectral properties (both in the region of d.c. and in terms of the overall
bandwidth), sufficient inherent timing for reliable clock extraction at the
receiver, and possibly some redundancy to provide error monitoring.
The degree of precision when equalizing for channel distortion may also
have to be considered, bearing in mind any self-equalizing properties
and the fact that multi-level codes require more precise equalization
than binary. For these reasons, many line codes designed to meet the
requirements of a variety of transmission systems have been invented.
Useful surveys, together with references to further details, are to be
found in Duc and Smith (1977), the *International Journal of Electronics*
(1983), and Bylanski and Ingram (1987). An interesting and quite
different approach using data scrambling and aimed at high bit rate
systems which currently require simple encoding and decoding
procedures is to be found in Fair *et al.* (1991).

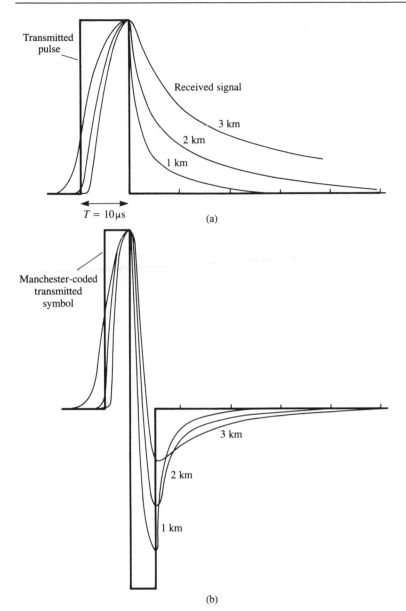

Figure 10.8 The pulse response of a metallic cable (0.4 mm copper pair): (a) received waveforms for a single transmitted pulse; (b) self-equalizing properties of the Manchester code

References

ANSI (1988) Basic access ISDN standard, T1.601–1988.

Brooks, R. M. and Jessop, A. (1983) 'Line coding for optical fibre systems', *International Journal of Electronics*, vol. 55, no. 1, 81–120.

Bylanski, P. and Ingram, D. G. W. (1987) *Digital Transmission Systems*, 2nd edn, Peter Peregrinus, London.

Catchpole, R. J. (1975) 'Efficient ternary transmission codes', *Electronics Letters*, vol. 11, 482–4.

Cattermole, K. W. (1983) 'Principles of digital line coding', *International Journal of Electronics*, vol. 55, no. 1, 3–33.

CCITT (1988) 'G.703: General aspects of interfaces', in *Blue Book*, ITU, Geneva.

Dawson, P. A. and Kitchen, J. A. (1986), 'TAT-8 supervisory system', *British Telecommunications Engineering Journal*, vol. 5, 153–7.

Duc, N. Q. and Smith, B. M. (1977) 'Line coding for digital data transmission', *Australia Telecommunications Research Journal*, vol. 11, 14–27.

Fair, I. J., Grover, W. D., Krzymien, W. A. and MacDonald, R. I. (1991) 'Guided scrambling: a new line coding technique for high bit rate fiber optic transmission systems', *IEEE Transactions on Communications*, vol. 39, 289–97.

Franaszek, P. A. (1968) 'Sequence-state coding for digital transmission', *Bell System Technical Journal*, vol. 47, 143–57.

International Journal of Electronics (1983) Special issue on line codes, vol. 55, no. 1.

Kawanishi, S., Yoshikai, N., Yamoda, S.-I. and Nakogawa, K. (1988) 'DmB1M code and its performance in a very high speed optical transmission system', *IEEE Transactions on Communications*, vol. 36, no. 8, 951–6.

Lender, A. (1966) 'Correlative level coding for binary-data transmission', *IEEE Spectrum*, vol. 54, 104–15.

Lender, A. (1981) 'Correlative (partial response) techniques', in *Digital Communications: Microwave Applications*, K. Feher (ed.), Prentice-Hall, Englewood Cliffs, New Jersey.

Sharland, A. J. and Stevenson, A. (1983) 'A simple in-service error detection scheme based on the statistical properties of line codes for optical fibre systems', *International Journal of Electronics*, vol. 55, no. 1, 141–58.

Waters, D. B. (1983) 'Line codes for metallic cable systems', *International Journal of Electronics*, vol. 55, no. 1, 159–69.

Yoshikai, N., Katagiri, K. and Ito, T. (1984) 'mB1C code and its performance in an optical communication system', *IEEE Transactions on Communications*, vol. 32, 163–9.

PART 3

Systems and networks

11 Digital networks—a review

GOFF HILL

11.1 Introduction

Although the original patents for pulse code modulated speech were established in the 1930s, PCM was not used for transmission until semiconductor technology allowed the technique, together with time-division multiplexing, to become commercially attractive around 1960. Since then digital transmission has been increasingly used in telecommunications networks around the world (Fig. 11.1). In the UK the earlier systems carried 24 telephony channels at a basic rate of 1.536 Mbit/s but by the early 1970s this was superseded in Europe by 32-channel systems conforming to CCITT recommendation G.732, which adopts a basic rate of 2.048 Mbit/s. USA and Japan continue to use the 1.536 Mbit/s rate as a basic standard, but each has developed different higher order standards. Since that time the progress of optical fibre techniques has been a key enabling technology in the development of systems with successively higher transmission rates (now extending to Gbit/s) which have also been standardized.

By the 1970s semiconductor technology had advanced to the point where digital exchanges (using stored program control, space and time switching, and common channel signalling between processors) became practicable and capable of providing a wider range of network services. The existence of digital transmission meant that analogue-to-digital conversions were reduced, allowing savings in cost and improvements in performance. By using large exchange processors to control a number of dispersed switches, considerable economies can be achieved for switching and for network administration. A typical implementation strategy for large national networks has been initially to introduce digital transmission to link local exchanges, then to extend digitalization into the trunk network (a digital backbone), followed by digital exchanges for trunk switching and later for local exchanges.

It has been possible for several years to carry signals in digital form from a local exchange across a network to a remote local exchange, but until recently, overall end-to-end digital transmission (from customer to customer) was only possible with a private circuit. The integrated services

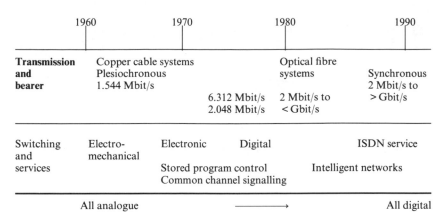

	1960	1970	1980	1990
Transmission and bearer	Copper cable systems Plesiochronous 1.544 Mbit/s	6.312 Mbit/s 2.048 Mbit/s	Optical fibre systems 2 Mbit/s to < Gbit/s	Synchronous 2 Mbit/s to > Gbit/s
Switching and services	Electro-mechanical	Electronic Digital Stored program control Common channel signalling	Digital	ISDN service Intelligent networks

All analogue ⟶ All digital

Figure 11.1   The evolution of digital systems

digital networks (ISDNs) (CCITT, 1988d) have provided digitalization for the final link to the user so that signals can be transmitted in wholly digital form from end to end across a network. This has been the last link of the chain to introduce end-to-end digitalization of national networks and the significance is that it extends the digital interface and signalling to the user to allow a wide range of new telecommunications services to be provided.

Earlier developments have provided immensely powerful capabilities for the modern network but they have also introduced some limitations in flexibility. Recent trends towards synchronous transmission techniques will more fully integrate the functions of transmission and route switching and will improve the flexibility of the transmission network, paving the way for multiservices. Looking further ahead, some of the limitations set by electronic technology are likely to be bypassed by more advanced optical networking techniques. The *telephony passive optical network* (TPON) (Stern *et al.*, 1987) appears particularly attractive to extend the use of optical fibre into the local network and to provide the basis for more widespread use and further development of the ISDN services, while optical multiplexing and switching techniques may offer improved traffic-handling capabilities for both local and core networks.

11.2 The telecommunications network

A telecommunications network is a highly complex arrangement of transmission, switching and management systems that provides a wide variety of services to its customers. In the UK alone there are over 25 million telephone lines linked by more than 2 million km of optical

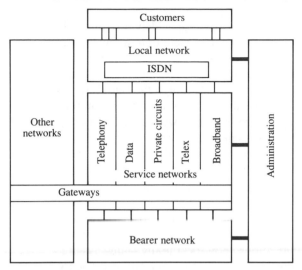

Figure 11.2 A typical telecommunications network

fibre as well as many multi-channel radio links and a vast copper cable network. A telecommunications network must be able to take signals from any one of its terminals, select a preferred route and transport the information accurately and reliably through the network to any required destination. To get the best out of the equipment that is installed, the network shares its resources whenever it can. For example, if a customer wants both a telephone line and a private circuit, the same cable and duct can be used to provide both services. Similarly, an exchange building can be used to house equipment for more than one type of service, but at the equipment level the services are generally kept separate at present.

The network may be considered in several functional parts as shown in Fig. 11.2. Customers require a range of services such as telephony, telex, data, etc., and these services are provided by service-specific networks which comprise principally switches, processors, signalling and transmission links. *Gateways* are provided so that interconnection can be made either between service networks or between the service networks and international or other operators' networks. Gateways between service networks are needed, for example, to allow a telephone modem terminal to transmit to a terminal on a data network.

The equipment for the service networks is located in exchange buildings and so a local network is needed to provide the link between the customers and the service networks. The local network provides individual links to each customer and is expensive to provide, accounting for some 40–50 per cent of total network costs. Because of the geographical distribution of terminals in a telecommunications network, the service networks must be widely distributed to achieve the most economic network structure. This is done by providing widely distributed local exchanges that typically limit the local line length to a few kilometres.

Because the service network equipment is housed in dispersed locations, a *bearer network* is needed to provide efficient bulk transport of signals between the network nodes. Although the signals for each service network vary at source, standard digital transmission formats allow capacity to be shared in the bearer network (see Chapter 6), raising the utilization of the transmission systems.

Apart from the equipment which provides the connection between terminals, it is necessary to monitor and control the status of the network so that it can be efficiently run and effectively maintained. This requires performance- and traffic-monitoring points throughout the network, and these must be linked to network management centres through an administration network. The administration network reaches all parts of a telecommunications network. Finally, gateways are provided so that networks can be linked together, for example to allow a telephony terminal to transmit data to a computer connected to a data network, or to allow interconnection of networks in a deregulated environment.

11.3 Local network

11.3.1 Copper cable networks

To a very large extent, today's local networks are analogue and operate over copper pair cables, although digital systems over optical fibre are being increasingly used, particularly to meet business needs. A typical copper-based local network structure is shown in Fig. 11.3 where both overhead and underground feeds are shown from a *distribution point* (DP) to the customers' premises. Fan out from the local exchange uses large copper cables, typically with 1000 or 2000 pairs. These cables need to be installed far in advance of demand so that both installation costs can be minimized and services can be provided rapidly on demand. As demand cannot be predicted accurately well in advance of when the cable is installed, primary and secondary cross-connect points are used to provide the flexibility to rearrange the lines to provide capacity where it is needed. Similarly, cables terminate on a *main distribution frame* (MDF) in an exchange so that equipment and cables can be connected as demand dictates.

11.3.2 Digital local links

Digital local links have been practical for major businesses for a number of years but their rate of penetration into the local network has increased since the recent introduction of ISDN services. The ISDN extends the digital interface of the service networks to the customer terminal and,

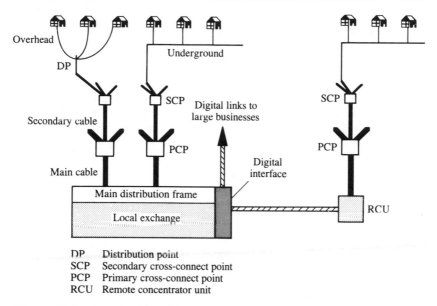

DP   Distribution point
SCP Secondary cross-connect point
PCP Primary cross-connect point
RCU Remote concentrator unit

Figure 11.3 Local network structure

through improved digital signalling, can provide a superior range of facilities. It is discussed more fully in Sec. 11.4.

For the future, passive optical fibre network techniques such as the TPON network show considerable promise for economic extension of the digital path to smaller customers. TPON uses a branching fibre network structure (Fig. 11.4) together with time-division multiple access to share a single transmitter, receiver, and fibre path at the local exchange end. A 2.048 Mbit/s TDM signal is transmitted from the local exchange to the customer and each customer is allocated one or more 64 kbit/s channels. For the return transmission a ranging protocol is used to synchronize bursts of data from customer terminals so that signals form a coherent TDM multiplex by the time they reach the exchange. Software addressing provides the flexibility needed to cope with physical changes to the customer base.

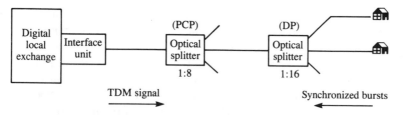

Figure 11.4 A typical TPON network structure

11.4 Service networks

11.4.1 Public switched telephone networks (PSTN)

A modern digital telephony network provides 64 kbit/s channels from end to end via *service switches* controlled by the user. *Operator services* to assist the user may be considered as part of the network as well as *intelligent network* facilities which are able to offer powerful service enhancements to the network. Although voice signals are most commonly carried, a network can also be used to carry a wide variety of signals, any in fact that can be made to conform to the 64 kbit/s format. This includes signals that are digital in origin, such as computer data, and those which are analogue, such as facsimile and telemonitoring, which must first be converted to digital form. Analogue signals will most commonly conform to the '4 kHz bandwidth' template required by the A/D converters designed for voice signals, but wider bandwidth signals may also be carried if they contain sufficient redundancy to allow bandwidth compression in the digital path prior to transmission over the 64 kbit/s sections. Still picture and even moving picture information can be carried in this way, as may be needed, for example, in a remote security monitoring system. A digital telephony network therefore is highly versatile, capable of both carrying a wide variety of signals between remote locations and providing remote paths as links between, or extensions to private, local area and metropolitan area networks.

Layered structure

A telephony network is a highly complex combination of hardware and software performing many different functions. It is useful to consider the major functions that the network performs as a stack of related layers (Fig. 11.5).

The transmission layer at the bottom of the stack provides for the transport of all signals between the switches and processors which, respectively, reside in the second and third layers. The switches control the routeing of all signals through the network and are themselves controlled by processors. The processors allow software control of the connection patterns of the various elements in the switching systems, operating on fast electronic switches in 'parallel' form to give fast call set-up and clear-down. This is good from the point of view of both the customer and the network operator whose equipment is idle for only very short periods of time as call set-up and clear-down processes take place. Software control removes the need for much of the manual interconnection and rearrangement work which was characteristic of earlier systems and leads to more rapid control of network operations.

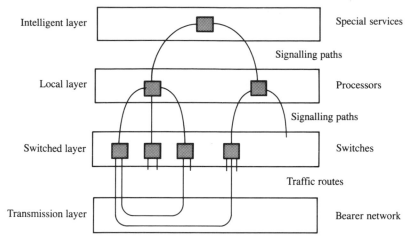

Intelligent layer — Special services
Signalling paths
Local layer — Processors
Signalling paths
Switched layer — Switches
Traffic routes
Transmission layer — Bearer network

Figure 11.5   Telephony network layers

Communications between the processors and between the processors and switches requires a powerful signalling system, such as the CCITT no. 7 system, as a means of controlling the various network functions. The top of the stack contains network intelligence, which allows a range of customized services to be provided in the network. The intelligence allows special network features to be set up such as customer networks with special billing arrangements.

Switching hierarchy

The switching layer itself is split into a number of hierarchic levels. The need for a hierarchy is illustrated by considering the interconnection of local exchanges in dispersed geographic areas (Fig. 11.6). A separate transmission route could be provided between each pair of exchanges to achieve the desired interconnectivity, leading to a total of $N(N - 1)/2$ routes for full interconnection (28 in this example). If the exchanges are widely dispersed the links will be long and there may be little traffic generated between them. It is more economical to share transmission systems by multiplexing where feasible and to concentrate the traffic. This can be achieved by introducing additional levels of switching. A single additional higher level switch (Fig. 11.6b) reduces the number of routes (to 8) and each will carry higher levels of traffic. The route lengths, however, may still be unacceptably long and of low capacity. This can be offset by increasing the number of higher level switches in a mesh structure (Fig. 11.6c). The number of links is increased (to 11) but their average length is reduced and the traffic in the higher transmission layer is increased. This arrangement is more likely to lead to an economic solution.

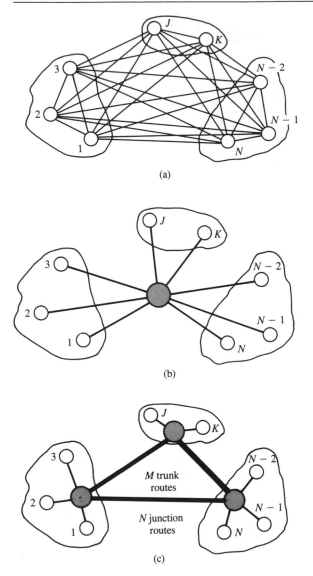

Figure 11.6   The need for hierarchic networks: (a) fully meshed network with $N(N - 1)/2$ routes; (b) star network with a higher switching level and N junction routes; (c) mesh network with a higher switching level

The mesh structure also provides the network with some resilience in the event of failure, as it is possible to provide more than one higher level route between any two local exchanges. For a very large network, additional levels of switching may be provided, as illustrated in Fig. 11.7. The design of a telephony network must take these factors into

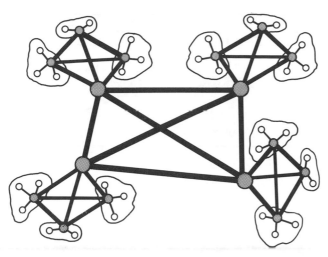

Figure 11.7   Layout of a large hierarchic network

account so that an optimum solution can be found for a given geographic area and flow of traffic.

Figure 11.8 is a representation of the BT telephony network. Here the level 1 switch (the *digital local exchange*, DLE) serves its own locality and provides direct access to operator services and to the level 2 switch (the *digital main switching unit*, DMSU) for longer distance calls. Customers may be connected directly to the level 1 switch or, in the case of rural customers, may be connected via a *remote concentrator unit* (RCU). It has always been expensive to provide service to rural customers because of the high cost of the line and because there are few customers to serve. A digital RCU allows a number of customers to share a small number of channels efficiently and, in addition, can simplify the interface at the local exchange by bypassing the main distribution frame (see Fig. 11.3). The DLE provides a full range of processor functions to control switches in the exchange and also to control the status of the dependent RCUs. By sharing a single DLE across a large number of RCUs considerable economies can be made in the cost of processing and network administration. The level 2 switches concentrate the trunk traffic and groom the signals on to preferred routes so that efficient use is made of the bearer network. At this level it is likely that the switches are either fully interconnected (as in the UK) or have a high degree of interconnectivity. It is at this level that international gateways are provided.

What has been said so far about the hierarchy is directly analogous to earlier analogue networks. The top level of the hierarchy, however, can use intelligent network techniques to provide customized services, and this depends heavily on a digital network. Examples of features offered by an intelligent network layer include special answering services, call prompting, and information services. The service providers often need

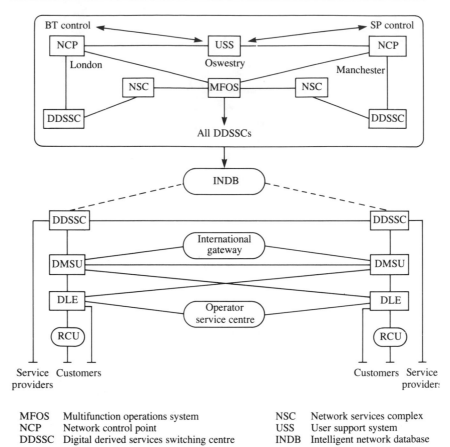

Figure 11.8 Telephony network hierarchy with two switching levels and intelligent services (from *British Telecommunications Engineering*, 1989, vol. 8, 513)

advanced features from the switch that conventional network users do not require (such as being able to vary the network routeing at different times) and so an additional type of switching centre can be provided at the top of the switching hierarchy containing these features. A smaller number of these special exchanges are needed than are required for trunk switching and so the software and services they provide can be shared by the trunk exchanges. Access to the service control features might be either via the switched network or via private circuits.

Signalling

A digital network relies on powerful signalling to ensure that complex management and control information between processors, switches, and

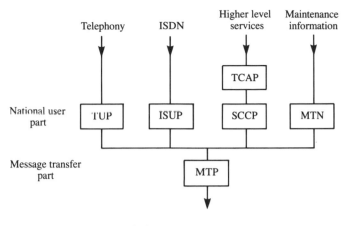

TUP   Telephone user part
ISUP   ISDN User part
SCCP   Signalling control connection part
TCAP   Transaction capabilities applications part

Figure 11.9 Application of signalling system no. 7 architecture

remote concentrator units can be relayed unambiguously between the network elements. The signalling may be carried in a special 64 kbit/s channel associated with the higher capacity (1.5 Mbit/s or 2 Mbit/s) blocks of the bearer network, which is described in Sec. 11.5. This is known as *common channel signalling*, as the signalling channel carries information for many calls in progress. An important advantage of common channel signalling is that user information and network signalling are separated and signalling can occur during the progress of a call. Several 64 kbit/s channels are normally dedicated to the signalling on each traffic route between two exchanges to carry the signalling information for all traffic on the given route. Often, more than one geographical path will be used so that in the event of line failure, for example, the signalling is not completely lost and the network can take restorative action and safeguard other calls in progress. Normally the signalling will be carried on the same route as the information channels but this is not essential. It may be more economical, for example, to provide a separate routeing for low traffic routes. Alternatively, separate routeing could be used to cope with failure conditions.

The CCITT no. 7 system can provide the necessary signalling capability using common channel techniques (CCITT, 1980; 1984; 1988c). This is a packet-like transmission system which uses 64 kbit/s channels and structures the signalling along the lines of the seven-layer OSI (open systems interconnection) model. The signal functions are layered into two main parts as shown in Fig. 11.9, a message transfer part (MTP) and a national user part (NUP). The MTP is common to all applications and provides information relating to the transfer of

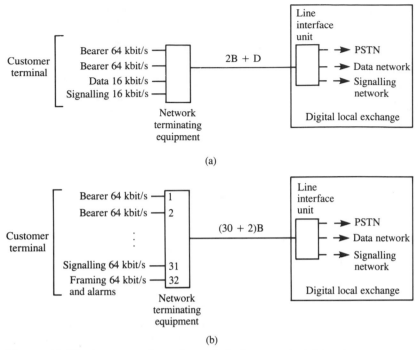

Figure 11.10   Integrated services digital network: (a) basic rate access; (b) primary rate access

information across the network. The NUP is application dependent and indicates the control conditions for specific calls (such as telephony or data). Additional blocks will be needed to cope with centralized network intelligence.

11.4.2 Integrated services digital networks

As the service and bearer networks have become progressively digitalized over the past 30 years, performance has improved and many new services have become available. It has been a logical step to extend the digital interface to the customers' premises and this is the purpose of ISDN. This has not been a straightforward task as a variety of customer terminal equipments are in use today as well as a variety of local transmission methods. To give maximum scope to the development of ISDN systems, a set of standard interfaces have been agreed in CCITT which assume that a network terminating equipment (NTE), sited at the customers' premises, will convert any transmission system on the network side or terminal equipment on the customer side to the appropriate interface.

Two principal ISDN options have now been standardized by CCITT, as illustrated in Fig. 11.10. A 144 kbit/s basic rate (BR) version offers two 64 kbit/s bearer (B) channels, and a 16 kbit/s data (D) channel is available for smaller customers. For higher channel requirements a primary rate (PR) version offers (30 + 2) bearer channels which can be used for speech or data; this rate is suitable for digital *private automatic branch exchanges* (PABXs). In PR systems two 64 kbit/s channels are reserved for signalling and synchronization. An alternative PR version is available based on 1.544 Mbit/s for the USA and Japan.

It has been possible for some time to provide a 64 kbit/s interface to a customer through the use of a private wire, so perhaps the most significant aspect of ISDN is that a powerful signalling technique is extended to the customer and it is through this that many additional services can be offered. In addition, the D channels provide new opportunities for enhanced services such as calling number identification and the ability to select service (such as telephony or data) on a call-by-call basis. Additionally, some features normally associated with a PABX can be provided.

11.4.3 Private circuit network

A *private circuit* (PC) network may have some similarity to a telephony network (using processor-controlled exchanges in a hierarchic manner) but does not provide the user with control over the switching. It is geared towards the business users who wish to set up or extend their own private networks. The switching in a PC exchange is under the control of the network operator and it is used to provide convenient and rapid set-up or rearrangement of circuits as demand requires. This type of switching is infrequent and is termed *cross-connect switching*. As there are fewer users of PCs than of telephones, fewer exchanges are needed in a PC network than in a PSTN. Consequently, exchanges are further apart and the 'local lines' longer. Often, users may be connected to the PC exchange via a 2 Mbit/s digital multiplexer sited at a convenient local PSTN exchange.

PCs at the higher rate of 1.5 Mbit/s and 2 Mbit/s may be routed directly over the bearer network (see Sec. 11.5). In Europe, cross-connect switches have not been developed for its 2 Mbit/s plesiochronous digital signals whereas in the USA 1.5 Mbit/s standard cross-connect switches have been developed and are in use. The alternative to using cross-connect switches is to use manual patch cords on flexibility frames to cross-connect 2 Mbit/s signals between systems. In Europe it is likely that cross-connect switches will be based on the recently introduced synchronous digital hierarchy standards.

11.4.4 Data networks

Telephony networks are based on *circuit switching* techniques in which a physical channel is set up and maintained for the duration of each call.

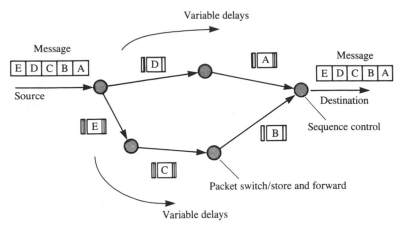

Figure 11.11   Principle of a packet switched network

Circuit switching matches both the signal and traffic requirements of speech channels since, when a call is set up there is generally a continuous exchange of information until the conversation is completed, then the call is cleared down and circuits and equipment in the network become available for the next call. In data communications the pattern is different as there can be long periods of time when a connection is set up but there is no exchange of information, then a memory or database downloads large volumes of data in a very short time. This type of 'bursty' traffic does not make efficient use of a physical channel and so data networks are based on *'packet' transmission and switching*.

Packet-switched networks break up each source message into relatively short blocks or packets for transmission through the network. Each packet is provided with a header containing a destination address and other control information. Packets are then relayed through the network in a *store and forward* fashion, each switch reading the address in the header to recover the routeing information (Fig. 11.11). Each packet has sufficient information associated with it to route it correctly through the network, but in 'connectionless' implementations (for LANs) the packets do not necessarily take the same route. This leads to the possibility of packets arriving out of order and hence the network must sort them out at the receive end.

Larger networks are 'connection oriented' where packets take the same predetermined route through the network. Connections between end terminals are not allocated a physical channel for the duration of a call; a connection is only provided when there is information to be transmitted. The channels are known as *virtual channels* and the network is referred to as a *virtual network*. In this way a large number of calls can share a single physical channel. This reduces the costs substantially but at the expense of additional control complexity compared to a circuit-switched network. Contention for switching channels is controlled by providing short-term storage facilities. The switch holds the packet until a suitable channel timeslot becomes free and then it retransmits it to the next node. This introduces a certain amount of delay in the transmission but for most data applications this can be tolerated. Where it cannot be tolerated (e.g., if real-time speech is to be transmitted) then it is possible to put information into the header to give it priority. In effect, a logical connection is set up across the network that emulates a physical circuit and the system must ensure the integrity, transparency, and packet sequencing across the network.

The X25 international standard (CCITT, 1980; 1984; 1988) provides the basis of public data and private networks, and can be used to illustrate these points. X25 has three levels of protocol. The first level is a physical level which defines the physical interface at the data terminal while the second frame level provides for error-free transport of packets. Frame handlers at this level add header information to define the sequence of packets. The third packet level defines the packet format and the control procedures that are to be used for exchanging packets. This level also controls the end-to-end signalling needed to establish the virtual connection.

Asynchronous transfer mode (ATM) is a further method of breaking a data stream down into small packets called cells. In ATM the packet lengths are uniform and a short header is used. When a call is set up, the routeing information is downloaded to the packet-switching nodes and the nodes allocate a short code to the transmitters which identify the particular call. The routeing information is stored as a look-up table so that when a packet is to be transmitted only the short code is included in the header, not the complete routeing address. The advantage of short packets is that they are less likely to experience long delays as they are relayed through the network and so avoid the long delays that make the transport of speech impracticable in X25. A packet length of 53 bytes has been agreed as the basis for international standardization which is suitable for LAN/MAN traffic.

11.5 The bearer network

The *bearer network* provides the transmission 'core' of a telecommunications network, providing the high-capacity capability to transport information between exchanges. Historically this has been referred to as the 'trunk and junction network', where the trunk part provides the long-distance capacity between trunk exchanges and the junction network provides links between local exchanges or from local exchanges to trunk exchanges. Interconnections between transmission systems and the links into telephony, telex, or other exchanges are made in repeater stations which are points of flexibility for the core network. Interconnections between transmission systems can be made at any level in the digital hierarchy according to the level of traffic.

11.5.1 Plesiochronous digital hierarchy

At the end of the 1980s the vast majority of transmission systems in use around the world were based on plesiochronous digital working, in which systems have the same nominal bit rate but can have a specified deviation. Plesiochronous systems provide primary rate multiplexing, which combines 64 Kbit/s channels into either a 1.536 or 2.048 Mbit/s frame and higher rate multiplexing which combines a number of primary rate signals. In forming the primary rate, bearer networks provide channel associated signalling whilst ISDN networks use a 64 Kbit/s common channel for signalling. Europe, which uses the 2.048 Mbit/s basic rate, and the USA and Japan, which use 1.536 Mbit/s, both use different methods to provide the signalling capability. The development of higher bit rate systems led to hierarchic standards known as plesiochronous digital hierarchy (PDH) which is described in Chapter 6. However, there is no agreed worldwide standard for PDH, as is witnessed by the different approaches to signalling and by the adoption of the two primary rates. The original formats for PDH made no provision for monitoring errors across the end to end path but path layer error monitor schemes were subsequently added during the 1970s and 80s.

PDH has provided a powerful vehicle for the introduction of digitalization, but it is now considered to have some significant limitation. Because the systems are not synchronous, when signals enter a transmission system, or pass from one system to another, there must be a process of bit-rate equalization (justification) in which additional bits are added into the signal stream until the bit rate just matches the system rate. After justification, the possibility of directly recovering, for example, a single 2.048 Mbit/s signal is lost, as the precise position of individual channels or bytes is not known, unless the signal is demultiplexed. This makes it difficult to extract individual channels and to use system capacity efficiently and, in practice, this leads to *multiplexer mountains* in which many channels are demultiplexed and remultiplexed many times as they traverse the network. Secondly, the PDH systems provide only limited possibilities for providing spare bits in the frame for operations, administration and maintenance (OAM). Thirdly, PDH does not offer standardized interfaces, making it generally impracticable to interconnect equipment from different suppliers.

11.5.2 Synchronous digital hierarchy

The problems of PDH have been addressed in the development of synchronous digital hierarchy (SDH), which has recognized the impact that synchronous techniques have had on 64 bit/s TDM switching. In a synchronous frame, the network has a priori knowledge of the position of each channel/byte. In the synchronous digital hierarchy justification is applied to each signal as it enters the network. This justification remains with the signal in a *container* until it leaves the network. A path overhead is added to the container to produce a *virtual container*. The overhead caters for packing and unpacking capacity blocks at the paths ends, performance monitoring, maintenance functions, and alarm status. Across a wide geographical area even a synchronous network may not be precisely synchronous because of the small timing differences that can occur in distributing a reference clock through secondary, tertiary, etc., standards. As a virtual container passes through a network from system to system its rate may also need to be finely adjusted. This second stage of justification (at byte level) modifies the virtual container to produce a tributary unit. The tributary unit is synchronous with the transmission frame that it sits in and the exact position of the virtual container in the frame is marked by a *pointer* carried in a separate part of the frame. The pointer allows the precise position of each virtual container to be identified and this feature makes it possible to perform time-switching operations to add/drop individual channels and to perform cross-connect switching. This has a major impact on signal management as it provides the possibility of selecting capacity in modular blocks and performing branching and cross-connect functions using digital methods without having to demultiplex all of the higher level signal.

SDH is defined by an international standard. The standard provides for higher capacity systems starting at the two basic rates of 1.544 Mbit/s and 2.048 Mbit/s and covers the hierarchic levels (CCITT, 1988, G.707), the frame structure (CCITT, 1988, G.708) and the multiplexing structure and overheads (CCITT, 1988, G.709). Generous provision is made in the overhead for OAM functions. Maintenance information can be associated with end-to-end customer routes as well as network sections and individual links.

In parallel with SDH developments, centrally controlled *synchronous digital cross-connect switches* (SDXCs) are being developed for control over topology and capacity routeing. SDXCs will provide flexibility between a number of different SDH line systems at the basic channel rate of 2 Mbit/s (1.5 Mbit/s in USA and Japan) and can be used to route blocks of $N \times 2$ Mbit/s signals through the network. They embrace the multiplexer function and can also remove the need for a flexibility frame between the exchange and the multiplexers. SDXCs will provide a valuable element in utilizing network resources in an effective and flexible manner.

11.5.3 Network reliability and protection

By careful choice in route planning, capacity between any two switching nodes in a network can be shared across physically separate routes, so that in the event of failure of one route (e.g., through a cable fault) a substantial proportion of traffic will be unaffected. The provision of spare capacity, switched in by protection systems, will also minimize the disruption to traffic in the event of failure. This can provide a capacity reserve for 'planned works'. In general, *manual changeover* systems provide a substantial improvement over no protection at all, but *automatic changeover* will give a more rapid restoration of service. Automatic systems can be provided on either a system or a network basis. System protection can provide the most rapid changeover as the decision to change over can be made locally. Network protection must refer the decision to an operations and maintenance centre computer and so takes slightly longer but it allows a greater degree of resource sharing. The introduction of digital cross-connect equipment not only allows new or rearranged routes to be provided in a network, but can also provide a protection facility, either on a system or a network basis.

11.6 Concluding comments

The main factors shaping the development of digital networks have been the following:

- Digital integrated circuit technology
- Digital transmission techniques
- Stored program control
- Digital switching techniques
- Powerful messaging
- Optical fibre technology

These factors developed relatively slowly at first, but when optical fibre systems began to be used from around 1980 they provided the last piece to the jigsaw. Since then, applications have developed rapidly and deployment has become widespread. Digital processing has allowed the development of a wide range of network services requiring different network characteristics, and optical fibre technology has had a profound effect on network operating costs.

In the future, these aspects will no doubt continue to be developed, but further major influences are likely to be the development of intelligent network techniques and the development of all aspects of service and network management operations. Intelligent network techniques will aid the development of an even greater range of services. Broadband techniques will be further enhanced for the private networks of business users, and the interaction of the telecommunications network and personal computers will be developed for both business and personal applications. All of these will place demands on transmission systems to carry increased levels of traffic, together with the associated signalling information, to manage the network properly. The new SDH standards, which make generous provision for network management and control overheads in the frame structure, will play a key role in these changes. In the following chapters the transmission aspects will be considered in more detail, and some insight will be provided into the factors that will shape future developments.

References

Bellamy, J. (1982) *Digital Telephony*, Wiley, New York.

Breuer, H. J. and Hellstrom, B. (1990) 'Synchronous transmission networks', *Ericsson Review*, vol. 67, no. 2, 60–71.

Chidgey, P. J. *et al.* (1991) 'The role for reconfigurable wavelength multiplexed networks and links in future optical networks', in *Photonics Switching Conference.*

CCITT (1980) Recommendation on X25, *Blue Book*, ITU, Geneva.

CCITT (1984) Recommendation on X25, *Blue Book*, ITU, Geneva.

CCITT (1988a) Recommendation on X25, *Blue Book*, ITU, Geneva.

CCITT (1988b) Recommendation on transmission, 'G.707, Synchronous digital hierarchy bit rates', 'G.708, Network node interfaces for synchronous digital hierarchy', 'G.709, Synchronous multiplexing structure', *Blue Book*, ITU, Geneva.

CCITT (1988c) Recommendation on signalling, Q700 Series, *Blue Book*, ITU, Geneva.

CCITT (1988d) Recommendation on ISDN, 'I.420, Network termination interface', 'I.441, Data link layer (D channel) protocol','Q.921, Data link layer (D channel) protocol—more than 1 logical link', 'I.451, Network layer protocol','Q.931, Network layer protocol', *Blue Book*, ITU, Geneva.

Held, G. (1992) *The Multiplexer Reference Manual*, Wiley, New York.

Stern, J. R., Ballance, J. W., Faulkner, D. W. *et al.* (1987) 'Passive optical local networks for telephony applications and beyond', *Electronics Letters*, vol. 23, no. 24, 1255–7.

Ward, K. (1989) 'The BT telecommunications network', *British Telecommunications Engineering*, supplement to vol. 8, part 1.

12 Optical transmission in telecommunication networks

MIKE O'MAHONY

This chapter considers the function and performance of optical transmission in the modern digital network. The history behind the development of optical digital transmission is briefly reviewed before considering where optical transmission is used in a network hierarchy. The components, transmission techniques, performance and limitations associated with optical transmission in current networks are then considered. Future optical transmission is discussed and its impact on network design is considered.

12.1 Introduction

In the period between 1960 and 1980, telecommunications transmission was predominantly over copper cable, with twisted pair in the local network and coaxial cable in the trunk (or toll) network. Microwave radio links were also installed to carry some long-haul traffic. Analogue transmission techniques were used, and this required very careful design and planning to achieve reasonable quality and performance.

In the period up to the end of the 1970s, however, the whole network was exhibiting a number of problems. Examples of these were:

- The analogue nature of transmission posed problems for ensuring high quality over long distances because of the accumulation of distortion and other impairments. The solution to this problem required the development of digital techniques.
- The demands for increased capacity in the trunk network necessitated the development of 140 Mbit/s and later 565 Mbit/s transmission systems. The high loss associated with coaxial cable, however, meant that regenerator spacings as low as 1 km were necessary. As the

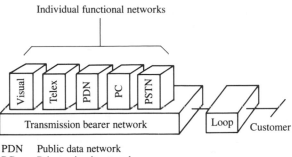

Individual functional networks

PDN   Public data network
PC   Private circuit network
PSTN Public switched telephone network

Figure 12.1   Structure of telecommunications network

average spacing between repeater stations was in the order of 50 km, this implied that a large number of buried regenerators were required, increasing cost and reducing reliability.

Against this background, from the mid-1970s optical transmission techniques were being researched and refined. The optical fibre medium (in its ultimate single-mode form) proved to be almost ideal. It combines virtually unlimited bandwidth with extremely low loss (0.25 dB/km) and so was ideally suited to the requirements of digital transmission. Semiconductor lasers were developed to access the fibre medium; however to this day their transfer characteristics are not particularly linear, although increasingly strenuous efforts are being made to design a linear device because of the requirements of cable TV. Thus laser sources are not particularly well suited to analogue modulation. Optical transmission and digital techniques proved, therefore, to be ideally suited, and the combination has proved exceptionally successful in providing high-quality transmission over very long distances. Currently, all the UK trunk network is optical and the penetration of fibre into the local loop is under consideration in most developed countries.

12.2 The telecommunications network

In the current telecommunications network, optical transmission is mainly confined to the bearer network, although there is increasing interest in the use of optical techniques in the local network. The bearer network is illustrated in Fig. 12.1 and essentially comprises all the high-speed transmission from the local exchange upwards (essentially the old trunk and junction system) (Ward, 1989). The bearer network supports all the functional networks, such as the public switched telephone network (PSTN) and the private circuit network (PCN). In the current plesiochronous network (see Chapter 6) the optical link is embedded in the bearer network and originates and terminates in a multiplexer and

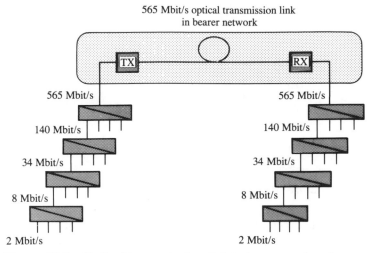

565 Mbit/s optical transmission link
in bearer network

Figure 12.2   Optical transmission link in bearer network

demultiplexer, as shown in Fig. 12.2. Signals arriving at the local exchange are time-multiplexed into 2–8–34–140–565 Mbit/s traffic streams which are then transported over the bearer network. At the far end an individual 2 Mbit/s stream is recovered by demultiplexing down to the 2 Mbit/s level. The need for this 'multiplex mountain' causes a number of problems, as it involves a large amount of electronics with associated reliability problems. The new synchronous network (Sec. 12.7) is designed to overcome this problem.

12.3 The optical link and its elements

Figure 12.2 is the schematic of an optical transmission link. It comprises a number of elements which together determine the end-to-end performance. In the following discussion, each of these elements is briefly considered.

12.3.1 The transmitter

The transmitter (TX) or source laser performs the electrical-to-optical signal conversion and is characterized by a transfer function of the form shown in Fig. 12.3(a) (Senior, 1985; Petermann, 1991). Optical oscillation occurs at the lasing threshold current, with the output increasing rapidly for currents exceeding this value. Intensity modulation, the main technique currently used, is achieved by modulating the bias current, with a typical efficiency of 1 mW optical output for 30 mA current. The transfer characteristic shown is idealized; in practice a degree of nonlinearity exists which makes the devices unsuitable for analogue use.

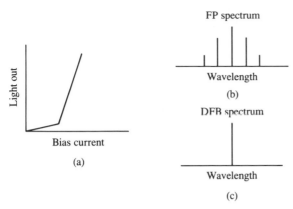

Figure 12.3   Laser characteristics

Currently, the most common wavelength used in the terrestrial network is 1.3 µm; for international undersea links the wavelength is generally 1.5 µm. The factors influencing the choice of wavelength are discussed later.

A key factor in link performance is the laser spectrum. As discussed below, this interacts with the fibre dispersion characteristic in a way that may significantly limit performance. Two types of laser are currently in use in the network and they are distinguished by their spectral profiles.

The *Fabry–Perot* (*FP*) *laser* has a spectrum which is extremely wide compared to that of the modulating signal. As illustrated in Fig. 12.3(b), the spectrum exhibits a comb of lines whose spacing is related to the laser cavity dimensions. Spectral widths in the order of 1–3 nm are typical. The majority of equipment in the present network uses these lasers as they are cheap and, until fairly recently, have satisfied the network requirements.

A major drawback of the FP laser is that it cannot be used at 1.5 µm, where fibre loss is lower than at 1.3 µm, because its wide spectral width causes significant penalties due to the interaction with the high fibre dispersion at 1.5 µm (see the following discussion on dispersion). The *distributed feedback* (*DFB*) *laser* was developed to overcome these problems.

The DFB laser has an integral filter within the laser structure which constrains the laser to operate at a single wavelength, as illustrated in Fig. 12.3(c). Because of the single-mode operation of the laser it can be used at 1.5 µm, where the fibre loss is low, but the dispersion is high. The main disadvantage is that of cost. At the time of writing, DFB lasers are significantly more expensive than FP lasers, mainly because they are more complex and require additional processing steps. For these reasons DFB lasers are used in undersea systems where a prime requirement is to maximize distance between repeaters. Terrestrial equipment using 1.5 μm DFB lasers for long-distance high bit rate applications is now also being deployed.

Although the static spectral spread of DFB lasers is very narrow (e.g., 10 MHz), under direct modulation spreading of the spectrum is found to occur. This wavelength 'chirp' is the result of the dynamic variation in refractive index that occurs when the bias current is suddenly changed by the direct modulation process. The coupling between photon flux and carrier density in a semiconductor is well understood and is defined in terms of the rate equations:

$$\frac{\mathrm{d}n}{\mathrm{d}t} = \frac{I}{eV} - \frac{n}{\tau} - g_0\Gamma\frac{n - n_\mathrm{t}}{1 + \varepsilon S} S$$

$$\frac{\mathrm{d}S}{\mathrm{d}t} = g_0\Gamma\frac{n - n_\mathrm{t}}{1 + \varepsilon S} S - \frac{S}{\tau} + \beta\frac{n}{\tau}$$

where: n = carrier density (n_t = threshold density)
S = photon density, Γ = confinement factor,
I = injection current, V = active volume,
g = material gain, τ = carrier lifetime,
β = spontaneous emission coefficient,
ε = gain compression factor.

For example, the second equation, which describes the photon density, is a function of the carrier density, n; as this varies so does the photon density, which in turn modifies the material refractive index. The variation in refractive index is translated into a variation in laser wavelength. Under current modulation, therefore, the narrow linewidth of a DFB laser is broadened; this broadening may be in the order of 0.1 nm (approximately 15 GHz) and has to be considered in any system design. Furthermore, as the wavelength variation is confined to transient parts of the modulating waveform, it is clear that its effect will be more noticeable at high bit rates when the pulse duration is short.

12.3.2 The fibre channel

Figure 12.4 shows the attenuation and dispersion characteristics of single-mode optical fibre (SMF), the type used in all modern telecommunications networks (Senior, 1985). This topic is covered in more detail in Chapter 2, Secs. 2.8 and 2.9.

The main features of the attenuation characteristic are as follows:

- Two main transmission windows are available at 1.3 μm and 1.5 μm, with corresponding attenuations of approximately 0.5 dB/km and 0.2 dB/km respectively.
- The lower limit to attenuation is set by Rayleigh scattering within the glass material.

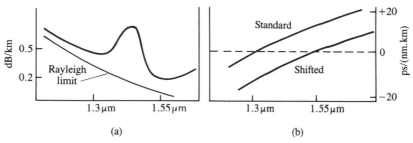

(a)   (b)

Figure 12.4 Optical fibre characteristics: (a) attenuation, (b) dispersion

- A water absorption peak occurs in the region of 1.4 μm, causing high attenuation, and so this region must be avoided.

As discussed above, fibre dispersion is a key feature in optical system design. The refractive index of the optical fibre is a function of wavelength, and this implies that energy at different wavelengths will travel at different velocities. Thus if a pulse is launched by a nonmonochromatic source, a spreading in the time domain will occur as the pulse traverses the fibre. The dispersion characteristics of the fibre arise from two sources: (1) the characteristics of the glass material itself, and (2) effects due to propagation within the glass waveguide (it should be clear that if the wavelength varies but the dimensions of the waveguide are constant, the boundary conditions will change). The effect of these two sources of dispersion is combined in the dispersion coefficient D (ps/(nm.km)), which is a measure of the temporal spread (in ps) experienced by the pulse emitted from a source of spectral width of 1 nm travelling through 1 km of fibre. Figure 12.4 shows that with SMF the dispersion has a zero in the region of 1.3 μm and a value of approximately 17 ps/(nm.km) at 1.55 μm.

A rule of thumb in system design is that 10 per cent spreading over the system length is acceptable, as this will cause a penalty of less than 1 dB. Thus a definition of the maximum distance L (km) that can be achieved for a given dispersion coefficient D, and bit duration T (ps) is:

$$L = \frac{0.1T}{D\Delta\lambda}$$

For example, with a bit rate of 2.5 Gbit/s, a wavelength of 1.55 μm and a source spectral spread of 0.05 nm, the maximum transmission distance is 48 km. At distances greater than this, dispersion will become a significant problem.

To overcome the limitation imposed by dispersion, special fibre has been designed (dispersion-shifted fibre, DSF), and installed in some European countries. By altering the waveguide structure of the fibre, the dispersion zero is shifted to 1.55μm, thus in theory low loss and low dispersion may be combined. In practice, the loss is a little greater than that possible with SMF, nevertheless it is currently being used in many undersea systems.

12.3.3 The receiver

The optical receiver (RX) performs the opto-electronic conversion and is a key element in the transmission link (Personick, 1973). The main elements in a direct detection optical receiver are a photodetector, a low noise preamplifier, and a pulse equalizer. For a bit error rate of 10^{-9} the signal-to-noise ratio at the equalizer output should be approximately 21 dB. The optical power incident on the photodetector required to achieve this is called the *receiver sensitivity*, an important parameter in system design.

In a properly designed optical system, the main noise source at the receiver output is that associated with the receiver electronics, thus careful design is needed to minimize this noise and to maximize detection sensitivity. A wide variety of receiver designs exists, for example, PIN or APD photodetectors combined with high impedance or transimpedance low-noise preamplifiers (Muoi, 1984). A typical sensitivity of a commercial receiver operating at 622 Mbit/s is − 35 dBm. As the bit rate increases the sensitivity decreases, generally by approximately 3 dB per octave.

More recently, receivers using optical preamplifiers to boost the received signal before photodetection have been developed. These are discussed in more detail in Sec. 12.5.2.

12.4 System configuration and performance

The preceding discussion looked at the main elements of an optical link. In this section transmission and system aspects are considered.

12.4.1 Modulation techniques

At present, most installed optical systems use baseband modulation, where the intensity of the laser source is modulated, either directly or through an external modulator. Direct detection is used at the receiver terminal to recover the signal. Intensity modulation, although simple (it is often likened to spark transmission), is very successful. Its main advantages lie in the simplicity of the method and its associated circuitry, and its excellent performance. Indeed, with the use of optical amplifiers to enhance detection sensitivity its performance is close to that achievable with coherent (carrier) detection. The disadvantage of baseband transmission is that the detector output is proportional to incident optical power, thus frequency and phase information is lost and detection efficiency is not ideal.

Transmission methods based on modulation of the amplitude, frequency or phase of the carrier are called 'coherent techniques' (see Chapter 4 and Linke and Gnauck, 1988). Although detection efficiency can be high, the receiver is generally very complex in comparison with direct detection receivers and practical implementation is difficult; for example, a local laser oscillator is required at the receiver, which must be tuned close to the frequency of the transmitter. In addition, the process of combining the received signal with the local oscillator is highly polarization dependent and polarization control or special polarization-independent receivers are needed. All of this adds to the complexity. Coherent detection, however, has a number of advantages:

- There is high detection sensitivity, close to the quantum limit
- As a result of the narrow linewidth associated with coherent transmission, it performs well where fibre dispersion is an issue. In addition, coherent detection enables dispersion equalization to take place in the electrical domain.

12.4.2 Choice of wavelength

As mentioned earlier, there are two main choices with respect to operating wavelength, namely 1.3 μm and 1.5 μm. The factors involved may be summarized as follows:

- *1.3 μm* Higher fibre loss; low dispersion on standard SMF; sources more readily available.

  The low dispersion means that cheaper Fabry–Perot lasers can be used on moderate-length systems without causing significant penalties. For these reasons, at present over 99 per cent of inland optical systems operate at 1.3 μm.
- *1.5 μm* Lower fibre loss; significant dispersion coefficient, unless dispersion-shifted fibre is used; special sources (e.g., DFB lasers) needed, which increases cost.

  High dispersion means that DFB sources or external modulators must be used, both of which are expensive. Low loss means a longer distance between repeaters where it is required. The wavelength is also suitable for erbium amplifiers. For these reasons 1.5 μm is used on most long-distance undersea systems.

12.4.3 Power budget

In planning an optical system a power budget is first calculated, to ensure that the link performance will meet the required standards. The power budget takes into account the transmitted power and the receiver sensitivity plus all the losses in the link. An example of a budget is as follows:

Transmitter launch power	-3 dBm
Receiver sensitivity (at 622 Mbit/s)	-35 dBm
Penalty due to dispersion	1 dB
System margin (for ageing, etc.)	6 dB
Allowable path loss	25 dB
Length (for 0.25 dB/km fibre loss)	100 km

In this case a maximum spacing between repeaters of 100 km is possible.

12.5 Future optical systems and networks

The discussion so far has addressed optical systems as they are currently used in the telecommunications network. In recent years, however, there have been many advances in optical transmission technology, which herald the deployment of advanced systems and networks. In this section some examples of the new technology are considered.

12.5.1 Optical amplifiers

Over a period of less than 10 years the optical amplifier has moved from a research curiosity to a commercially available component (O'Mahony, 1988). The amplifier offers a number of advantages, for example:

- Direct optical amplification with a minimum of electronics.
- Terahertz bandwidth which closely matches that available in a fibre. The matching of fibre and amplifier allows the development of 'transparent' systems which can operate with any bit rate or transmission format.
- When designed from semiconductor material it can be integrated to form functional components such as optical switches and wavelength convertors.

In particular the ability to offer gain bandwidth products in the order of 30 dB . THz is unmatched by any electronic component.

Optical amplifiers have a wide range of possible applications. Some of the major ones are:

1 *Optical gain block* For example, to overcome splitter losses in fibre distribution systems.

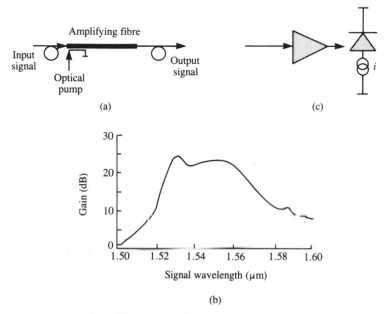

Figure 12.5 The fibre optical amplifier

2 *Transmitter power booster* For example, to increase the amount of power launched into a system fibre and hence increase the maximum unrepeatered length.

3 *Linear amplifier repeater* This will probably be the first commercial use of the amplifier. It allows the replacement of a complex opto-electronic regenerator with restricted bandwidth by a simple all-optical component. In addition, the use of a linear amplifier repeater enables the deployment of wavelength-division multiplex transmission, as the amplifier can amplify a large number of channels simultaneously with a minimum of crosstalk.

4 *Optical receiver preamplifier* Positioning an optical amplifier immediately in front of an optical receiver enables the sensitivity of the receiver to be enhanced. By this means the sensitivity of a simple direct detection system can approach that of the more complex coherent detection receiver.

Historically, the initial studies of optical amplifiers focused on semiconductor material, however in the late 1980s attention focused on amplifiers based on optical fibre. The compatibility of the amplifier fibre with the system fibre gives many advantages, for example minimum coupling losses, and it is this component that is now the main focus of attention.

Figure 12.5(a) is the schematic of an erbium fibre amplifier operating at 1.5 µm (Atkins *et al.*, 1989). The amplifier comprises the following parts:

1   An input fibre coupler which combines an optical pump with the input signal. As the signal and pump are at different wavelengths it is possible to design the coupler to have close to zero loss in each path.

2   An optical pump to provide the required population inversion in the amplifying fibre. This is generally a semiconductor laser, and currently two possible pumping wavelengths are under consideration, 980 nm and 1480 nm; the former is preferable from a purely technical point of view (e.g., it gives higher efficiency in terms of dB gain/mW pump—6 dB/mW as opposed to 2.5 dB/mW), however there are unresolved questions concerning pump laser reliability. The pump power required depends on a number of factors, however, by way of example, gains in the order of 30 dB are possible with less than 20 mW.

3   An amplifying fibre. This is typically a silica fibre doped with the rare earth erbium and codoped with germania and alumina. The length of the fibre may vary from 10 to 30 m. The pump energy is absorbed at wavelengths of 980 nm and 1480 nm, resulting in population inversion of the medium, and signal amplification is obtained centred about 1.536 µm. The latter wavelength is fixed by virtue of the rare earth used and it is not possible to operate this amplifier in the 1.3 µm band (of course considerable research activity is centred around the development of such components).

Figure 12.5(b) is an example of a typical gain versus wavelength characterisitc of an erbium amplifier. The main feature is that gains in the order of 25 dB are possible with bandwidths of approximately 35 nm.

A crucial feature of optical amplifiers is their noise performance. At the output of the amplifying fibre a number of noise components exist (Mukai *et al.*, 1982). These can be defined with reference to Fig. 12.5(c), which shows an optical amplifier followed by a photodiode. The amplifier noise is assessed by evaluating the currents flowing in the diode due to the various noise components. Likewise, the signal level in the diode can be determined and this enables an assessment of the signal-to-noise ratio to be established.

Assuming P = instantaneous input signal power, G = amplifier gain, R = diode responsivity (ampere/watt), B = electrical bandwidth, e = electronic charge, hv = photon energy, γ = population inversion parameter, Δf = optical bandwidth, then the mean current components flowing through the diode are:

$$I_0 = GPR + \gamma R[G - 1]\Delta fh\nu$$

where the first term is the amplified signal and the second term is the amplified spontaneous emission.

To determine the signal-to-noise ratio requires an evaluation of the noise variance currents. The optical amplifier contributes four noise components, two related to the shot noise associated with the currents caused by the signal and amplified spontaneous emission, and two beat noise components caused by interaction between spontaneously emitted photons and signal and spontaneous photons. As the bandwidth of the amplifier is very large (35 nm), and that of the signal comparatively small, an optical filter can be used after the amplifier to reduce the significance of a number of these components. In addition, if the amplifier gain is high, the effects of electronic noise are negligible. With these conditions it can then be shown that the dominant noise component is the beat noise between signal and spontaneously emitted photons, and the noise variance is expressed as:

$$i^2 = 4eR\gamma G(G - 1)PB$$

The signal-to-noise ratio (SNR), for the case $G \gg 1$, is then given as:

$$\mathrm{SNR} = \frac{RP}{4e\gamma B} = \frac{P}{2h\nu B}\frac{\eta}{2\gamma}$$

The extreme right-hand expression is written in a form to allow comparison with the ideal shot noise limit. The latter is obtained by considering a signal with power P incident on a diode with responsivity $R = e\eta/h\nu$ (where η is the quantum efficiency, h Planck's constant and ν the optical frequency).

Assuming Poisson statistics with the variance equal to the mean:

$$\mathrm{SNR}_{\text{shot noise limit}} = \frac{\eta P}{2h\nu B}$$

Thus optical preamplification enables detection to within a factor 2γ of the ideal shot noise limit where γ is generally close to unity for fibre amplifiers. Thus the sensitivity of an optical preamplifier receiver can approach within 3 dB of the shot noise limit.

12.6 Optical amplifier systems

The development of optical amplifiers has coincided with the need for changes in telecommunications networks. There is an increased demand for capacity, flexibility, and reliability. Optical amplifiers are seen as offering potentially transparent systems which can accept virtually any

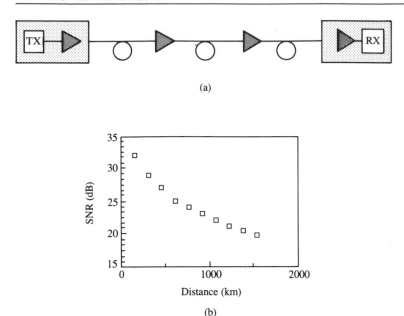

(a)

(b)

Figure 12.6   Optical amplifier system

signal format and are not restricted by electronics and hence can be upgraded more gracefully. Optical amplifier technology is inherently simple and should be more reliable than conventional opto-electronic regenerators. A flexible system design can be adopted using a transmitter booster, linear amplifier repeaters, and optical preamplifiers. This enables the development of increased span unrepeatered systems as well as unregenerated long-haul systems.

Amplifier systems, however, are analogue in nature and limitations may arise from a number of sources, for example, accumulated noise, dispersion, and fibre nonlinearities, thus careful design is necessary.

Figure 12.6(a) shows a general system design incorporating a transmitter booster, a cascade of amplifier repeaters, and a receiver preamplifier. It is instructive to examine some of the main features of this system. In all system designs two aspects need to be studied, signal and noise performance.

Signal design

The transmitter booster enables an increase in the launched power. Current commercial amplifiers give an output power of approximately $+15\,dBm$, with an input power of $0\,dBm$. At the receiver an optical preamplifier can enhance the detection sensitivity. By way of example, a state-of-the-art $2.5\,Gbit/s$ preamplifier receiver has been demonstrated to have a sensitivity of $-43\,dBm$, about $8\,dB$ improvement on the basic detector. Thus at 2.5 Gbit/s, without repeaters, a total fibre loss of 58 dB could be tolerated (without any allowance for margins), approximately 18 dB improvement on an unamplified system. This is very significant for undersea systems. However, the effects of dispersion must also be considered.

In a repeatered system, an additional problem is encountered in the signal path. Each amplifier generates its own amplified spontaneous emission, and the noise accumulates along the amplifier chain. If the level is too high it can cause amplifier saturation with a consequent reduction in gain. The total amplified spontaneous emission (ASE) after N amplifiers is:

$$P_{ASE}(N) = N\gamma(G - 1)\Delta fh\nu$$

Calculation shows that after 20 repeaters with 30 dB gain (and $\gamma = 1$) the total power is approximately $+10$ dBm, this is likely to cause problems.

Noise design

The noise variance components also accumulate along the amplifier chain, with the signal-spontaneous beat noise increasing as N, for example. Thus the signal-to-noise ratio will decrease as the number increases. Figure 12.6(b) shows the results in the case where the amplifier gain has been set at 25 dB, which with a fibre loss of 0.25 dB/km allows a 100 km distance between repeaters. In this example a total distance of approximately 1500 km is possible (equivalent to 15 amplifiers). To increase this distance optical filtering is required to further reduce the noise.

12.6.1 Multiwavelength systems

The concept of *wavelength-division multiplex* (WDM) has been around for a number of years. The principles are described with reference to Fig. 12.7(a). A number of signals at different wavelengths are combined by a multiplexer on to a single fibre. At the receive terminal a demultiplexer recovers the original channels for further processing. Thus the technique allows the fibre to be shared between a number of signals, improving its utilization.

There are a number of reasons, however, why this technique has not been extensively exploited. In the initial period of fibre deployment there was an over-capacity of fibre and there was little incentive to improve utilization. A more significant reason is that new components were necessary (multiplexers and demultiplexers) which introduced additional losses. For example, with a moderate number of wavelengths (e.g., four)

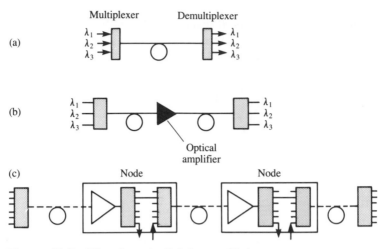

Figure 12.7 Wavelength-division multiplex

the minimum insertion loss is in the order of 5 dB, yielding a total end-to-end insertion loss of 10 dB. In addition, at each regenerator node the individual wavelengths had to be recovered, each individually regenerated and then recombined for onward transmission. In a fibre-rich environment therefore, WDM had little to offer for simple point-to-point transmission.

The development of optical amplifiers, however, could radically change the economics of WDM. Owing to their large bandwidth, a number of channels of different wavelengths can be simultaneously amplified without the need for demultiplexing (Fig. 12.7b). In the case of the fibre amplifier, crosstalk is negligible and demonstrations of the simultaneous amplification of 100 channels have been reported. With optical amplifiers, therefore, wavelength-division multiplexing may provide the key to the development of flexible optical systems which can be reconfigured in capacity as growth demands.

It is the networking potential of WDM, however, that is of greatest interest. Figure 12.7(c) shows the possible evolution of a point-to-point amplifier system into a network. Six wavelengths are routed through a number of nodes, which may represent exchanges in the network. At each node an amplifier is included to overcome losses associated with the fibre and the multiplexer and demultiplexer. Within the node the wavelengths are demultiplexed and, if required, information can be dropped off locally on a selected wavelength. In the same way, locally generated information can be inserted for transmission. This provides the add/drop function required of a network, as discussed in Chapter 6. In the following section we conclude by looking at how this concept might be incorporated into the future synchronous network.

12.7 The synchronous network

Over the next decade telecommunications networks in Europe, Japan, and the USA will move from plesiochronous to synchronous operation. There are a great many reasons for implementing changes, some of these are:

- Improved control and management facilities will provide faster customer response and better use of capacity.
- Improved availability and quality of service through reduced hardware and automatic protection.
- Reduced costs through the availability of hardware standards.
- Future expansion through software enhancements.

Key new hardware elements to be introduced in the synchronous network are:

- The add/drop multiplexer, which replaces the conventional multiplexer and allows direct access to 2 Mbit/s in an 155 Mbit/s stream.
- The HACE (higher order cross-connect equipment) switches in the long-distance (upper levels) of the network. These allow direct cross-connection at the 2 Mbit/s level. In operation these switches would be controlled by a central computer to enable remote reconfiguring of the network.

12.7.1 Problems with growth

The synchronous network incorporates optical transmission in its conventional role as point-to-point links between the nodes. Over the next 10 to 15 years however, a number of factors may arise that could influence the penetration of optics in this network, moving the optics from simple point-to-point links to performing functions within the nodes themselves. Some of these factors are:

- Growth in the network through increased demand for telephony; that for new wideband services may place great pressure on the cross-connect switches in the upper layer of the network. As currently envisaged they switch all the traffic entering a node.
- New services may demand wide bandwidth routes through the network.

One proposed solution to these problems is an optical layer extension to the telecommunication network hierarchy (European Commission RACE Project, 1992). The concept is illustrated in Fig. 12.8(a). The upper layer of the proposed synchronous network may comprise 20–100 nodes. At each of these nodes the majority of traffic entering the node

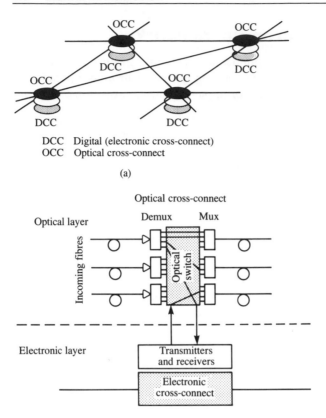

DCC   Digital (electronic cross-connect)
OCC   Optical cross-connect

(a)

Figure 12.8   Optical layer (RACE project)

will go straight through and only a small amount will be required locally. However, in the electronic implementation all the traffic is routed through the cross-connect switch. The optical layer concept (Fig. 12.8b) combines an optical space switch with the electronic switch; the optical switches are interconnected in an optical layer. In this implementation the through traffic entering the node remains in the optical domain and is switched (optically) through by the optical switch; no electronic conversion is encountered. Local traffic is switched through from the optical into the electronic layer. Thus optics is used where it is best to manipulate large capacities and electronics where it is best to carry out the fine-grain processing.

Switching and routeing are achieved through a combination of wavelength and space selection. Each incoming fibre bears a number of wavelengths which are demultiplexed within the node. The space switch then allows an individual wavelength on an individual fibre to be selected for through transmission or for local processing.

The above discussion illustrates one possible application of advanced optical techniques to the developing telecommunications network. Another application is the area of the local network, where fibre to the home has been growing in importance in recent times. Indeed some countries, notably Japan, have made a commitment to a firm programme of fibre installation. The deployment of fibre local networks for telephony is occurring at a time when there is great interest in developing fibre networks for cable TV distribution. These areas represent the main challenge for optical transmission over the next decade.

References

Atkins, C., Massicott, J., Armitage, J. *et al.* (1989) 'A high gain, broad spectral bandwidth erbium doped fibre amplifier pumped near 1.5 µm', *Electronics Letters*, vol. 25, no. 14, 910–11.

European Commission RACE project (1992) 'Multi wavelength transport network', January 1992–1995.

Linke, R. A. and Gnauck, A. H. (1988) 'High capacity coherent lightwave systems', *J. Lightwave Technology*, vol. 6, 1250–69.

Mukai, T., Yamamoto, Y. and Kimura, T. (1982) 'S/N and error rate performance in AlGAs semiconductor laser preamplifier and linear repeater systems', *IEEE Trans. Microwave Theory Tech.*, vol. MTT-30, no. 10, 1548–56.

Muoi, T. V. (1984) 'Receiver design for high-speed optical fibre systems', *J. Lightwave Technology*, vol. LT-2, no. 3, 243–67.

O'Mahony, M. J. (1988) 'Semiconductor laser amplifiers for future fibre systems', *J. Lightwave Technology*, vol. 6, no. 4, 531–44.

Personick, S. D. (1973) 'Receiver design for digital fibre optic communication systems', *Bell System Technical Journal*, vol. 52, no. 6. 843–86.

Petermann, K. (1991) *Laser Diode Modulation and Noise*, Chapter 3, Kluwer, Boston.

Senior, J. (1985) *Optical Fiber Communications*, Chapter 6, Prentice-Hall, Englewood Cliffs, New Jersey.

Ward, K. (1989) 'The BT telecommunication network', *British Telecommunication Engineering Journal* (supplement), vol. 8, part 1, 1–31.

13 Mobile radio systems—design considerations

CHARLES HUGHES

13.1 Introduction

Mobile systems differ considerably from other digital systems because they depend on free-space radio transmission over paths that are constantly changing. In common with other radio systems, they require the only commodity in the field of telecommunications that is in limited supply, namely the radio spectrum. More importantly, the vagaries of radio propagation are such that the radio links cannot be designed in absolute terms as can, for example, optical fibre systems or even fixed microwave links. When the mobile stations are moving about within a service area, it is only possible to design the system in probabilistic terms so that an adequate service is provided for most of the users and for most of the time.

Until relatively recently, mobile systems tended to be used only in situations where communication was necessary while on the move. Their main use was therefore confined to 'private' users such as the emergency services and industries involving dispersed and mobile units. The users were often experts, in the sense that they had had some training in passing stereotyped messages, so that the system fulfilled its main function. A 'public' service was provided in some countries, but it tended to be a minor subsidiary to the main, wired network. The reasons for the relative unpopularity of mobile service were not hard to find. The transmission was poor because of the radio propagation path, the service was not available in many areas and, because it was a minority service with few subscribers, it tended to be expensive. Operating agencies were not overanxious to develop the service because of the limited amount of radio spectrum that was available.

In the 1980s and 1990s a rapid development in mobile radio systems took place. Advances in semiconductor technology meant that inexpensive equipment could be constructed for operation at higher frequencies in the UHF bands, and hitherto unused parts of the spectrum became available for mobile systems. Other technological advances such as frequency synthesizers permitted the introduction of cellular systems. These not only overcame the spectrum restrictions by permitting frequency reuse at much shorter distances, but also provided much better transmission quality. The high initial investment required to provide the infrastructure for cellular systems could be justified only if a relatively large number of users could be found within a short time, but both government agencies and entrepreneurs saw the opportunity and provided the necessary capital. This in turn provided the incentive for the manufacturing industry to exploit emerging technologies and produce the low-cost transceivers necessary for a popular service.

Even today, digital transmission over mobile links is mainly restricted to the control signalling and some low-speed data over voice links. However, the situation is again changing rapidly. New techniques for error detection and correction are being developed to overcome many of the deficiencies of the radio transmission path, and advances in speech coding enable digital systems to compete with analogue in terms of the bandwidth required. Large-scale integration has enabled the hardware necessary for the more complex equipment to be manufactured at low cost. In the longer term, digital mobile systems may predominate over their analogue counterparts for both public and private networks.

In many cases the mobile networks have been established in an almost 'green field' situation without any external constraints, except those of radio spectrum, with few interfaces to fixed networks. These have provided a unique opportunity for system designers, in that they have been able to design a system from scratch without having to make any of the compromises that are usually necessary to make a new system compatible with existing networks. The aims of this chapter are therefore twofold. Firstly, the application of digital techniques covered in the preceding chapters will be described in relation to mobile systems. Secondly, mobile systems will be used to explain some of the approaches used in designing a major system that is based on a wide range of basic techniques and technologies.

13.1.1 Types of system

The International Telecommunication Union (ITU) recognizes land, maritime, and aeronautical mobile services (ITU, 1987). The mobile satellite service is divided into the same three categories. In this chapter, we shall be concerned with civil land mobile systems.

Mobile systems may be classified in terms of the service offered:

- *Paging* Transmission of brief messages from the fixed station to a mobile.
- *Net* Mobiles communicate with a dispatcher at the base station.

- *Switched* (*trunked*) Communication on any one of a group of channels.
- *Extensions to a fixed network* (cordless telephones).

13.1.2 Modes of operation

A further subdivision may be made in the mode of operation of the system although the terminology used is sometimes misleading:

- *Simplex* Radio paging systems are truly simplex since the transmission is in one direction only, but the term is also applied to 'push to talk' systems operating on a single frequency.
- *Semi-duplex* (also known as two-frequency simplex) The mobile and base transmitters operate on different frequencies but not simultaneously.
- *Full duplex* Provides two-way operation throughout a 'call'; the mobile link can then be used as a normal telephone connection.

13.1.3 Frequencies

Mobile systems have had to compete with other users for a share of the radio spectrum, but mobile services have a unique claim in that the only way of communicating with a mobile unit involves the use of free-space electromagnetic waves. Frequency bands in most parts of the radio spectrum have been allocated for mobile operation in the ITU table of frequency allocations (ITU, 1982), but the increasing need for mobile communications has meant that demand has exceeded the allocation of spectrum. The ideal frequency for mobile operation appears to be in the region of 300 MHz. Much lower frequencies are subject to greater noise levels and long-range interference, while at much higher frequencies obstruction by buildings causes increased propagation losses. Most 'private' and older 'public' systems use the VHF bands. Increasing demand for mobile services, coupled with the falling cost of higher frequency RF components, has led to greater use of UHF bands. Modern 'cellular' systems and extensions to fixed networks ('cordless' telephones) now use frequencies in the region of 900 MHz. As services continue to expand, new frequency bands (e.g., 1700 MHz) are coming into use and proposals have been made for use of EHF bands. Frequencies in the region of 60 GHz may be used for short-distance mobile communications. The oxygen in the atmosphere increases the attenuation (CCIR, 1986), which prevents their use for longer distances but reduces interference from other transmitters on the same frequency.

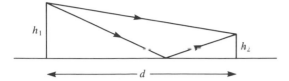

Figure 13.1   Ray paths for line-of-sight propagation

13.2 Mobile radio propagation

The radio path between a mobile and a base station represents the most difficult part of mobile system design because of the uncertain nature of the transmission path. Although considerable research has been devoted to means of estimating the radio path loss in probabilistic terms, it is still something of a hit-or-miss affair. For any major system, sample field-strength measurements are usually made to supplement the design calculations and, even then, changes have often to be carried out after a system has been put into operation. It is, of course, important to consider propagation effects not only within the service area but also at greater distances because of interference with other systems using the same frequencies.

13.2.1 Distance effects

Although the free-space attenuation of radio waves follows an inverse square law, earth reflections come into play when the mobile antenna is only one or two metres above the ground. The ray paths for a base station to mobile link are shown in Fig. 13.1.

When transmission takes place over a perfectly conducting earth, the reflected wave causes destructive interference because of the phase reversal on reflection. The received signal power is then proportional to:

$$\frac{h_1 h_2}{d^4}$$

In practice, because the earth is not a perfect conductor, the received signal power follows a $1/d^k$ law, where k is usually in the range 3.4 to 3.8 (Matthews, 1965; Griffiths, 1987).

13.2.2 Effects of obstructions

Except when propagation takes place over flat countryside or desert, clear line-of-sight paths between the mobile and the base station are comparatively rare. Hills and large buildings obstruct the radio waves and reduce considerably the received signal strength. However,

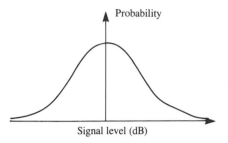

Figure 13.2 Distribution of 'local mean' signal strength

diffraction effects result in some energy being scattered into areas in the shadows of such obstructions. Several semi-empirical techniques have been devised to permit the estimation of the field strength of the signals over a comparatively large area at a given distance from the transmitter. This is known as the *area mean field strength.* For our purposes of system design, it will be sufficient to assume that the radio path attenuation is equal to the loss calculated on the basis of a $1/d^{3.6}$ (say) law plus an additional loss sometimes known as the *clutter factor*.

The additional attenuation due to the clutter factor is not the whole story. The effects of the radio shadows cast by buildings in particular are not constant, so the signal undergoes fading as the mobile moves through the radio field. This is generally referred to as *slow fading* and takes place over distances of the same order as the dimensions of buildings, usually some tens of metres. When the signal level is measured *in decibels*, the probability distribution approximates to a normal distribution as shown in Fig. 13.2. The distribution is often referred to as a *log–normal distribution* and this type of slow fading as *log–normal fading.* The standard deviation is usually about 5–8 dB.

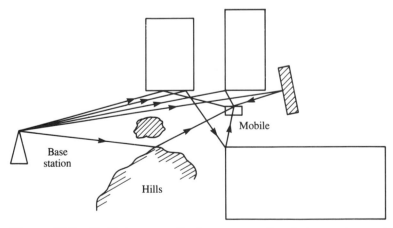

Figure 13.3 Multiple ray paths for propagation between base station and mobile

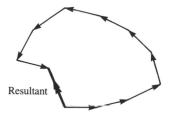

Figure 13.4   Phasor diagram for waves received over multiple ray paths

13.2.3 Effects of multipaths

Because of irregular diffractions and the effects of reflections from buildings and other obstructions, a signal received at the mobile may be regarded as the sum of the waves arriving over several different ray paths (Fig. 13.3). Because of the differing path lengths, these rays have differing amplitudes and phase angles on arrival at the antenna. The resultant amplitude of the received signal is the sum of the phasors representing the individual rays as shown in Fig. 13.4. As the vehicle moves the ray paths change by differing amounts and the resultant varies in amplitude and in phase. This is known as *Rayleigh fading* or *rapid fading*, so called because the changes take place over distances of less than a wavelength.

If the components of the ray phasors are resolved along two arbitrary axes at right angles it may be shown that the resultant of the components along either of the axes follows a binomial distribution. For a large number of rays, the resultant in phase space has a two-dimensional normal distribution (Fig. 13.5) (Cattermole, 1986).

The probability that the resultant amplitude will lie between r and $r + \delta r$

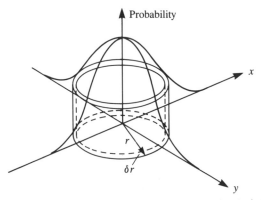

Figure 13.5   Two-dimensional amplitude/phase probability distribution for a large number of phasors with random phase angles

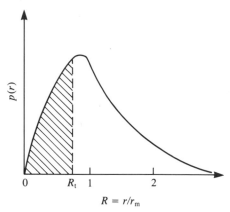

Figure 13.6 Rayleigh probability distribution

is proportional to the volume of the cylinder of thickness δr shown in Fig. 13.5. The probability distribution for the amplitude is then:

$$p(r) = \frac{r}{r_{\mathrm{m}}/2} \exp\left(-\frac{r^2}{r_m^2}\right) \tag{13.1}$$

Where r_{m} is the r.m.s. amplitude of the combined waves. This is known as the Rayleigh distribution and is plotted in Fig. 13.6.

The corresponding cumulative distribution function is

$$p(R \leqslant R_{\mathrm{t}}) = 1 - \exp\left(-R_{\mathrm{t}}^2\right) \tag{13.2}$$

where R and R_{t} are the amplitudes measured relative to the r.m.s. value.

13.2.4 Overall fading

It is useful to summarize the overall effects of fading of the received signal as shown in Fig. 13.7(a). In addition to the distance effects, the absence of a line-of-sight path, particularly in built-up areas, gives rise to an additional fixed attenuation known as the *urban clutter factor*. When these two effects are taken into consideration, the result is known as the *area mean* signal strength. Variations in the effects of obstructions give rise to slow fading but the *local mean* signal may be regarded as constant over a distance of the order of 10 metres. Multipath effects then give rise to rapid fading about the local mean.

13.2.5 Fading rate

In digital systems, we are often concerned with the signals fading below a given threshold level. In Fig. 13.7(b) the proportion of time the signal spends below the threshold level, R_{t}, is clearly the sum of the individual

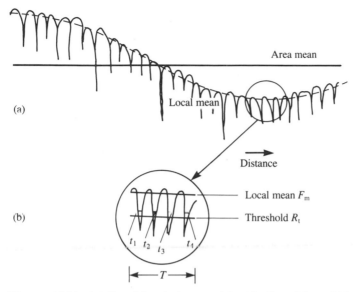

Figure 13.7 (a) Combined slow and fast fading (b) rapid fading about a local mean level

fades $t_1, t_2, t_3, \ldots$ divided by the total time of observation, T. The proportion of time spent below the threshold may be found from the cumulative Rayleigh distribution equation (13.2).

However, this is not the whole story. In digital systems, it is important to know whether the errors which are associated with fades occur in short or long bursts. It will be seen from Fig. 13.7(b) that the probability of the signal being below the threshold can be the same irrespective of whether there are many short excursions below R_t or a smaller number of longer ones. It is therefore important to find the total number of times the signal level falls below the threshold during the interval T. This is known as the *fading rate*. It has been shown by Rice (1948) that the number of upwards (or downwards) crossings of the threshold per wavelength is given by:

$$N(R_t) = \sqrt{2\pi} R_c \exp(-R_t^2) \qquad (13.3)$$

The fading rate, that is, the number of upwards crossings of the threshold per second can then be found by multiplying the above expression by V/λ, where V is the velocity of the mobile in metres per second and λ is the wavelength.

13.2.6 Radio propagation and system design

The propagation phenomena described above make the design of a mobile system somewhat different from the design of, say, a microwave point-to-point link. In particular, the depth of fades encountered on a

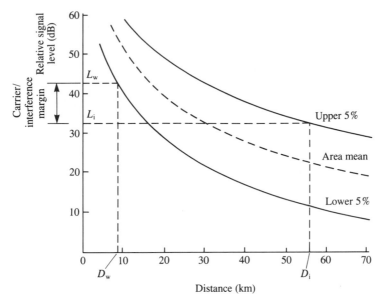

Figure 13.8 Signal level conditions for the determination of reuse distances

mobile link are much more severe than for a fixed link. The rate of fading can also vary over a much wider range, from zero for a hand-held mobile or a vehicle stationary in traffic to several hundred fades per second for a mobile on a high-speed train. Apart from fading of the wanted signal, cochannel interference caused by another transmitter operating on the same frequency is also subject to wide variations in strength and this must be considered in the system design.

Frequency reuse

The rapidly increasing demand for mobile radio systems means that there is pressure to use those parts of the radio spectrum allocated to the land mobile service as efficiently as possible. This means that the transmitter power should be restricted to that necessary to give satisfactory reception over the service area. The same frequency channel can then be used by another mobile operator in an area which is not too distant from that of the first operator. This concept of minimum reuse distance over a country (or ideally over a continent) then gives maximum spectral efficiency.

In an idealized system, the area mean received signal strength as a function of distance may be calculated from Eq. (13.1) and this is plotted in Fig. 13.8. Slow or log-normal fading then gives a spread to the area mean value and, as explained earlier, the level variations, measured in decibels, exhibit a normal probability distribution. It would be impracticable to design the system for a given level of wanted signal-to-interference ratio and to ensure that the system never operated under worse conditions. Instead, we take some arbitrary low percentage (say 5 per cent) and design the system so that the wanted signal only falls below a given level (L_w) for 5 per cent of the time and the interfering signal only exceeds a second given level (L_i) for 5 per cent of the time. The difference between the two levels ($L_w - L_i$) then represents the margin of wanted signal to interference.

In Fig. 13.8, it is assumed that the standard deviation for the log-normal fading distribution is 6 dB, a typical value for an urban area. From tables of the area under a normal distribution curve, it is found that 5 per cent of the area is contained under that part of the curve lying outside 1.65 times the standard deviation and these limits are plotted in Fig. 13.8. Suppose, for example, we require a margin of wanted signal-to-interference ratio of at least 9 dB for satisfactory operation. If the service area were 8 kilometres in radius (D_w), the frequency assignment plan should ensure that a second base station operating at the same frequency with the same antenna height and transmitter power is spaced at a distance (D_i) at least 56 kilometres from the mobile (i.e., 64 kilometres from the first base station). Of course, for part of the time, even the 9 dB margin will not be achieved and the service will appear unsatisfactory to the user. However, since it is only a small fraction of the time, and these conditions only occur when the mobile is near the periphery of the service area in the direction of the interfering base station, the overall performance may well be regarded as satisfactory. In this example, we have considered only transmission from the base station to the mobile, but a consideration of the geometry shows that the reuse distance for the mobiles is very similar.

In practice, the situation is rather more complex. Rapid or Rayleigh fading has not been considered and an additional margin of signal to interference is necessary to allow for such effects. Service areas are not usually circular in shape and shadows within a given area may lead to the necessity for repeater stations. Transmitter powers and antenna heights are not always the same and this leads to further complications in frequency planning. Nevertheless, the techniques outlined are used as a basis for frequency assignment planning by the regulatory authority.

Cellular systems

These are designed to give very efficient frequency reuse by employing many low-powered base stations. Each base station provides service over a comparatively small area (a *cell*), which may be as small as a 1 km radius in city centres, where there is a high density of mobiles. Larger cells, up to about a 30 km radius, are used in rural areas. As the mobile moves from one cell to another, its frequency is changed automatically to that of the new cell, a process known as *handover* (US—handoff). Within a large city, a given frequency can then be reused many times (MacDonald, 1979). This leads to a high degree of spectral efficiency, which is further increased since the frequency assignments and base station locations can all be planned in relation to the expected traffic as part of the overall system design.

Noise

As in any telecommunications system, the effects of noise should be considered as part of the system design. In most mobile systems, manmade noise, particularly from ignition systems, tends to predominate but even this becomes less serious as the higher frequency bands come into use (CCIR, 1986). In most short-range land mobile systems sufficient transmitter power can be provided to overcome the effects of noise. Most systems in countries which have a high mobile penetration are then limited by interference from transmitters operating on the same frequency, and the effects of manmade noise become a secondary consideration.

13.3 Modulation and coding

13.3.1 Analogue modulation

Until recently, mobile radio systems have used analogue modulation for voice transmission. Because of the deep and sometimes rapid fading encountered, it has been difficult to achieve a satisfactory design of receiver automatic gain control (AGC) which would permit the use of amplitude modulation. Frequency modulation has therefore been preferred. Only a modest amount of AGC is required to prevent overloading of the RF and IF stages of the receiver and a limiter is employed to restrict the signal amplitude before it is fed to the discriminator. Frequency modulation also has the advantage of a 'capture' effect. If the carrier-to-interference ratio is above about 10 dB, then the effects of interference are substantially reduced.

Unfortunately, for frequency modulation to be effective, the frequency deviation has to be much greater than the highest modulating frequency and this leads to excessive use of the radio spectrum. More recently, attention has been paid to the use of single-sideband amplitude modulation with an audio tone transmitted in the middle of the speech band used to control a rapid-acting AGC circuit at the receiver (McGeehan and Bateman, 1984). The narrow bandwidth required for single-sideband transmission appears to offer high spectral efficiency, but since the system is linear, it is very susceptible to interference. The reuse distance is consequently reduced and this counteracts to some extent the effects of reduced bandwidth.

Data transmission

Even on systems designed primarily for voice, data sometimes has to be carried. In switched (trunked) systems in which several mobiles share a group of frequencies, signalling information has to be carried to allocate the channels to mobiles and for other control functions. In cellular systems, additional information has to be transmitted to control the handover process. Combinations of audio frequency tones may be used to transmit the signalling information or, alternatively, digital modulation of a subcarrier may be used. In very large systems, separate radio frequency channels are used to carry the signalling information.

Where user data is carried on an 'analogue' system, a subcarrier is often used in the same way as a modem is used over a fixed telephone circuit.

13.3.2 Digital modulation

Although digital modulation in the form of PCM is widely used for voice transmission over fixed circuits, it has not in the past found favour for mobile application because of the wide bandwidth required. More recently, advances in voice coding have reduced the bandwidth required, and a reduced susceptibility to interference has permitted smaller reuse distances compared to analogue FM.

Voice coding

The digital coding of voice signals has undergone considerable development in recent years. If the objective is merely to transmit distinguishable words, a very low bit rate can be achieved, but most users require a degree of naturalness and recognition of the speaker. The most promising coding techniques at the present time appear to be the category of analogue/digital converters known as *linear predictive coders* (Atal, 1982; Jayant, 1986). The general principle is to transmit a model of the complex acoustic filter formed by the vocal tract together with a model of the excitation formed by the breath. If the model is updated every 20 milliseconds or so, good 'toll quality' speech can be achieved by a coder/decoder combination operating at about 12–16 kbit/s. This may be compared with the 64 kbit/s standard for PCM. Codes giving acceptable quality speech at even lower bit rates may be developed in the near future.

A further advantage of digital speech coding is that once the signals are in digital form, error correction techniques, discussed in Sec. 13.4, can be used to produce a decoded voice signal that is virtually indistinguishable from that achieved over a fixed channel, provided the 'raw' error rate is sufficiently low for the correction system to operate

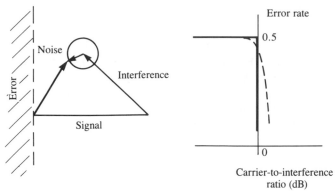

Figure 13.9   Reception of FSK signals

effectively. If the raw error rate is too high for the correction system to cope, the decoded speech becomes completely incoherent.

Digital keying

Frequency-shift keying (FSK) is normally used to modulate the transmitters. The pulses are often shaped (e.g., Gaussian minimal shift keying, GMSK) (Pathsupathy, 1979) to restrict the bandwidth of the emission. It is then possible to restrict the bandwidth occupied by the transmitted signal so that it is approximately numerically equal to the bit rate of the keying signals.

For reception by a frequency/phase discriminator, it is only necessary to distinguish between a digital 1 and a 0 as shown in the phasor diagram, Fig. 13.9. The resultant received signal is the vector sum of the phasors representing the wanted signal (a transmitted 1), the interfering signal, and noise. Then, assuming a perfect detector, if the resultant lies to the right of the broken line, a 1 will be received correctly. Only if the resultant lies to the left of the broken line will an error occur.

Under ideal conditions and in the absence of noise, the error rate is zero provided the level of the wanted carrier exceeds that of the interfering signal. Below that level, the error rate rises very rapidly. The reception of a signal in the presence of random noise rather than an interfering signal is discussed in Chapter 1 but the general approach is the same. The effect of noise in this case is to make the transition from zero error rate to high error rate more gradual as shown but, under typical working conditions, the noise-free case represents a fair approximation and may be used to analyse the effects of fading on the error performance.

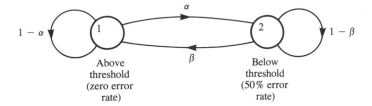

Figure 13.10   Markov model of digital fading signal

Digital fading model

As shown above, the output of the frequency discriminator in the receiver may be regarded as being error-free when the received signal level is above that of the interference and having a very high error rate (about 50 per cent since with binary keying there is always an even chance that a digit will be received correctly) when it falls below the interference threshold. Under fading conditions, the operation may be represented by a discrete two-state Markov model (Fig. 13.10) (Gilbert, 1960).

In the 1 state, the wanted signal is above the level of the interfering signal and the error rate is zero. During a fade, the wanted signal falls below the threshold set by the interfering signal and the error rate rises to 50 per cent. Transitions between the two states can only take place at discrete intervals corresponding to the bit-rate of the signals. If the probabilities of transitions between the two states are α and β as shown, then the probabilities of remaining in the states for a further bit interval are respectively $(1 - \alpha)$ and $(1 - \beta)$. The model thus simulates the effects of a period of error-free reception followed by bursts of errors. The probability (P_2) of the model being in state 2 may be found from the cumulative Rayleigh distribution, Eq. (13.2). The fading rate, $N(R_t)$, may also be found from Eq. (13.3) where the threshold level is that of the interfering signal. If b is the bit rate of the signal, the expected number of fades in the bit interval is:

$$N(R_t)/b \tag{13.4}$$

This is equal to the combined probability that the signal is below the threshold and that it will rise above the threshold during the bit interval, that is

$$P_2\beta$$

Applying the conditions for statistical equilibrium enables the transitional probabilities to be calculated:

$$\alpha = \frac{N(R_t)}{b(1 - P_2)} \tag{13.5a}$$

and

$$\beta = \frac{N(R_t)}{bP_2} \tag{13.5b}$$

The probability distribution for the durations of the error bursts (i.e., the periods when the error rate is very high) then takes the form of a geometrical distribution with a mean duration $1/\beta$.

13.4 Error control

Because of the variable nature of the transmission path, some form of error control is almost invariably required for digital operation.

13.4.1 Error coding

Error control coding is discussed in depth in Chapter 9. The following are the main error control techniques in use in mobile systems. Two or more of them may be used together to overcome the effects of poor transmission.

- *Multiple transmission/majority decision* Each data block is sent an odd number of times and corresponding bits in each block are compared; a majority decision is used to determine whether a given bit is more likely to be 0 or 1. This is a very simple approach and involves little delay but it is very wasteful of transmission capacity. It is therefore only used in control signalling when a very short message has to be transmitted in a very short time (for example, a handover command in a cellular system). In some cases the message may be transmitted 11 times.
- *Automatic repetition (ARQ)* (Comroe and Costello, 1984) This involves use of simple error detecting code (e.g., cyclic code) and, if an error is detected, a request for a repetition of the faulty block is sent on the return channel. Long delays are involved, so this approach is not suitable for control signalling. The possibility of the repetition request being in error leads to additional complications.
- *Block error correcting codes* These provide correction for a small number of errors in the block and detection of a greater number of errors. When used on their own, they are not very effective for

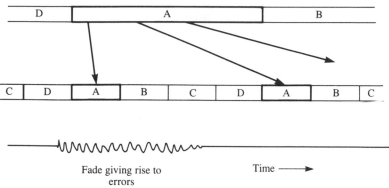

Figure 13.11 Interleaving data blocks to spread the effects of error bursts

dealing with high error rates but they require only a modest increase in transmission capacity.

- *Convolution coding* This provides a very effective method for overcoming the effects of high error rates, particularly if only short bursts are involved. The transmitted bit rate must be increased by a factor of two or three but the degree of error correction achieved is very much greater than for a simple majority decision system.

13.4.2 Interleaving

If a mobile is moving very slowly, the effects of multipath fading may cause it to remain in a deep fade for a relatively long interval. The result will be a long burst of errors even though the local mean signal level is much higher than that of the interfering signal. In this case, even powerful error correcting techniques will be insufficient to prevent errors in the output. One way to alleviate the problem is to interleave part of a data or speech block with parts of other blocks. For example, a data block A may be divided into four and interleaved with parts of data blocks B, C, and D as shown in Fig. 13.11. When an error burst that would destroy most of data block A occurs, it only affects one quarter of each of the four blocks. A powerful error-correcting technique such as convolutional coding might then correct the errors in all four blocks and given an error-free output. Of course if the vehicle were travelling even more slowly, the error burst might be so long that even the convolutional decoder could not cope. A higher degree of interleaving, say dividing a data block into eight and interleaving with seven other blocks, might overcome the problem, but this would increase the time delay involved in transmitting a data block. The system design therefore involves estimating the trade-offs between error correction capability and transmission delay.

13.4.3 Data transmission on analogue systems

As in the fixed network, the increasing penetration of mobile services has resulted in a demand for data transmission over the speech path. There is no problem in providing the basic modem as indicated in Sec. 13.3.2 but there is, as yet, no agreed standard for the error correction. A typical data transmission facility designed for operation over the voice channels of a cellular mobile system has the following characteristics.

- Modem   V26 bis
- Modulation of subcarrier   QPSK/FSK
- Bit rate   2400/150 bit/s
- Forward error correction   (16,8) punctured BCH or (72,68) Reed–Solomon code
- Automatic repetition   Based on cyclic redundancy check
- Interleaving depth   4

13.5 Time-division multiplexing

If digital transmission is to be used for radio links, an obvious further step is to consider the possibility of time-division multiplex (TDM) operation, as discussed in Chapter 6. At first glance, there may not appear to be any advantage in imposing an additional multiplexing technique, since the channel selection (tuning) of radio transmitters and receivers already offers a method of frequency-division multiplexing. Indeed, there is a disadvantage in time multiplexing, since unlike a fixed system, the transmission channels from the mobiles are not available at one point (the MUX terminal) for arrangement into sequential timeslots. Radio propagation delays make it difficult for a mobile to decide exactly when to transmit its timeslot or 'burst', and guard intervals have to be left at the end of each burst to prevent overlapping when successive bursts arrive at the base station. However, there are advantages in using TDM for mobile links. If a continuous two-way link is required as on a normal telephone circuit, transmission and reception have to take place simultaneously. If there is only one antenna at the mobile (the usual case) a *diplexer* filter has to be provided to prevent the transmitter output from interfering with the receiver frequency. Even if the transmit and receive frequencies differ by several megahertz, a complex filter is needed to prevent the transmitter output, which may be some 10s of watts, from affecting the receiver designed to accept signals of a few microvolts. However, if the signals are time-multiplexed, the signal bursts can be phased so that transmission and reception do not occur at the same time, and a simple switch prevents the mobile transmitter from affecting its receiver. Other advantages include the use of a single transmitter for several channels at the base station rather than one

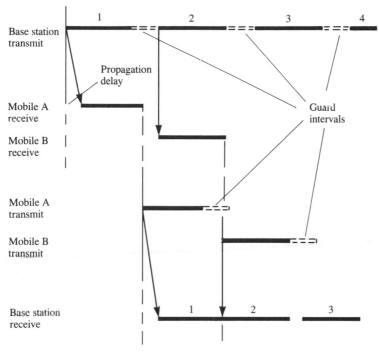

(Note: mobile A on periphery of cell;
mobile B close to base station)

Figure 13.12 Timing for multiplex bursts

transmitter per channel. Furthermore, it is a relatively simple matter to switch off a transmitter during the silent periods between talkspurts and so reduce some of the effects of cochannel interference.

13.5.1 Multiplex timing

The timing for the multiplexed channels is shown in Fig. 13.12. The master control is exercised by the base station and each mobile times its own transmission burst to take place a fixed interval after reception of a burst from the base station. For example, it is assumed that mobile A is on the periphery of the cell and mobile B is close to the base station. The burst from the base station to mobile A will thus take a relatively long interval to propagate and if A times its return burst to follow immediately afterwards, it will arrive back at the base station. The bursts to and from mobile B will take a very short time to propagate and it will be seen from Fig. 13.12 that if a guard interval were not left after each transmission burst, there would be an overlap between the bursts from the two mobiles as received at the base station.

13.6 Design of the multiplex burst

Although the design of the multiplex burst may appear to be relatively straightforward, it represents a good example of system design, since it involves the interaction of several different technological fields. The objective is to determine the length and composition of a burst which will provide effective speech and data transmission combined with efficient use of the radio spectrum. The following example is somewhat simplified but it will serve to illustrate the design process.

13.6.1 Design requirements

The following are assumed to be the main requirements for the design:

- Number of channels in multiplex   8 (the minimum to justify the additional costs of multiplexing)
- Maximum cell radius (D_1)   35 kilometres (to give a sufficiently high traffic level in rural areas)
- Frequency   900 MHz (wavelength, $\lambda = 0.333$ metres)
- Maximum vehicle speed (V_m)   250 km/h or 69.4 m/s (to include mobiles on high-speed trains)
- Maximum coding delay   Approx. 20 milliseconds (to avoid adding unduly to delays within the fixed network which may involve satellite links)
- Maximum delay spread (Δ_m)   10 microseconds (in mountainous regions)
- Bandwidth   Not to exceed 200 kHz, corresponding to 25 kHz per channel (the current spacing for analogue FM systems in Europe)

13.6.2 Design sequence

The sequence shown in Fig. 13.13 represents a greatly simplified version of the process adopted. In practice, several iterations would be necessary to reach a satisfactory definition.

Speech coder

It is assumed that satisfactory speech coding can be achieved by a form of linear predictive coder operating at 12 kbit/s.

Speech and data block formation

If the coding delay is restricted to about 20 milliseconds, this means that it is convenient to form the encoded speech into blocks corresponding to 20 milliseconds of speech, i.e. 240 bits. Data at speeds up to 12 kbit/s

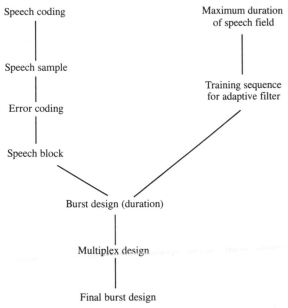

Figure 13.13 Steps in the design of a multiplex burst

may also be assembled into similarly sized blocks. The 240-bit block can then be treated as the unit for the application of error correction.

Error correction

As indicated in Sec. 13.4.1, convolutional coding provides a powerful method of forward error protection. It is assumed that a half-rate convolution code is used. The number of bits in a protected block is thus doubled and four bits must be added to allow the final bits to pass through the shift register of the error coder. The protected block size is thus increased to 488 bits.

Bandwidth

Under the above conditions, the minimum bit rate for the eight-channel system is:

$$8 \times 488 \times \frac{1000}{20} = 195\,200 \text{ bit/s} \tag{13.6}$$

In fact, the bit rate is likely to be appreciably greater than this, as will be seen later. If the upper limit of 25 kHz per channel is to be met, then signals in excess of 200 kbit/s will have to be carried within a maximum bandwidth of 200 kHz. This can only be achieved by the use of adaptive equalization (see Chapter 3).

Duration of burst

The problem in using adaptive equalization over mobile links is that the training signal used to set the coefficients on the transversal filter are dependent on the transmission path. This means that a new training signal must be sent each time the mobile moves a distance sufficient to cause an appreciable change in the path characteristics. Then, assuming a training signal is associated with each burst, the maximum duration of a burst will be constrained by the time for a fast-moving mobile to change its position sufficiently for a change to be required in the equalizer coefficients. Note that the restriction is one of time rather than of the number of bits. This explains one advantage of multiplexing, since the bit rate is then increased so that a larger number of bits may be transmitted before a new training signal is required. The proportional overhead of the training signal is therefore reduced.

A rough criterion for the maximum permissible transmission time following a training signal may be established by considering the phasor diagram of Fig. 13.4. If none of the phasors changes by more than a small angle (say, $\pi/10$), then the phase and amplitude of the resultant will not be changed by very much. On this criterion, the maximum transmission time after the training signal is the time required by a fast-moving mobile to travel a distance corresponding to a change in the phase angle of the radio wave of $\pi/10$, i.e., 1/20th of a wavelength. At 900 MHz, the wavelength is 0.333 metres, so that:

$$\text{Maximum transmission duration} = \frac{\lambda}{20 V_m}$$

$$= \frac{0.333}{20 \times 69.4} = 0.24 \text{ ms} \qquad (13.7)$$

In fact, it is possible to take greater advantage of the training signal by transmitting 0.24 ms of speech or data before the training signal and the same amount after it. At the receiver the first part of the burst is delayed in analogue form until the adaptive filter coefficients have been set.

Guard interval

For a cell of radius 35 km, the maximum propagation delay is:

$$\frac{2D_1}{c} = \frac{2 \times 35 \times 10^3}{3 \times 10^8} = 0.233 \text{ ms} \qquad (13.8)$$

where c is the velocity of light (3×10^8 m/s). (The factor of two in Eq. (13.8) arises because the mobile takes the timing of its transmitted burst from the received burst.)

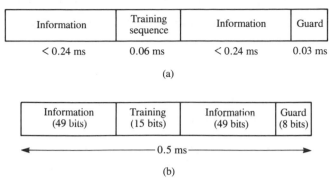

Figure 13.14   Field durations and numbers of bits in transmitted burst: (a) preliminary timing estimates; (b) final design

If a guard interval of this duration were inserted after each burst, it would be comparable with the time available for transmission of the information and would be very wasteful of capacity. However, only the first burst when the mobile applies for access to the system need have a guard interval as long as this. The base station can estimate the approximate transmission delay from that first burst and thereafter can instruct the mobile to adjust the timing of its transmitted bursts. If the guard interval is restricted to about 0.03 ms, mobiles moving on a radial path at the maximum speed (69.4 m/s) will have to update the burst transmission time at the following time intervals:

$$\frac{3 \times 10^8 \times 0.03}{69.4 \times 10^3} = 130 \text{ s} \tag{13.9}$$

This is about the time of an average telephone call, so only on rare occasions will the burst transmission time have to be changed during a call.

Training signal

In the design of an adaptive equalizer for a dispersive path plus a filter whose bandwidth is equal to the bit rate, a rule of thumb is that the number of taps on the equalizer should be equal to six times the number of bits transmitted in the maximum dispersal time ($\Delta_m = 0.01$ ms). For this system, the duration of the training signal will then be approximately 0.06 ms.

Fitting blocks to bursts

Figure 13.14(a) shows the approximate durations of the fields within a transmission burst. These are maximum durations, and the next step is to fit a coded block into a convenient number of bursts together with the training signal and the guard interval.

The maximum duration of the burst is approximately 0.57 ms and the corresponding total time for the transmission of eight bursts from the different channels (i.e., one *frame*) is about 4.6 ms. A 20 ms speech block can then be conveniently transmitted in five bursts, the time interval between the start of successive bursts in the same channel being 4 ms. For eight time-divided channels, burst duration, including the guard interval, becomes 0.5 ms. Since the speech block contains 488 bits, this means that each burst should contain 98 information bits. It will be seen from Fig. 13.14(b) that the minimum number of bits required for the training signal is:

$$\frac{0.06}{0.2} \times 49 = 14.7 \tag{13.10}$$

which on rounding up becomes 15.

The minimum number of bits for the guard interval is:

$$\frac{0.03}{0.2} \times 49 = 7.35 \quad \text{(8 on rounding up)} \tag{13.11}$$

The number of bits in each field of a burst is then as shown in Fig. 13.14(b).

Keying speed

From Fig. 13.14(b), it will be seen that 121 ($2 \times 49 + 15 + 8$) bits have to be transmitted in one burst interval of 0.5 ms. The keying speed for the multiplex is then 242 kbit/s.

13.6.3 Topics involved in the transmitted burst design

The compromise design for the relatively simple burst/multiplex design has involved:

- Speech coding
- Radio fading
- Modulation
- Radio spectrum utilization
- Propagation delay
- Traffic speed distribution
- Signal detection
- Adaptive equalization
- Error coding

13.6.4 Pan-European cellular system (GSM)

A digital cellular system designed to permit operation throughout
Europe has been defined and is now coming into operation. It is
generally referred to as the GSM system (GSM comes from the *Groupe
Speciale Mobile* committee set up by the CEPT to define the system).
The burst design previously described is loosely based on this system,
but it has been simplified to illustrate the design process. Apart from the
slightly different requirements specification and design parameters, the
main differences lie in the control signalling. This is carried out by using
a multiframe arrangement, that is, by devoting every 13th frame to
signalling. This increases the keying speed still further. Handover
instructions require a very rapid response and are signalled in a voice or
data traffic burst. A flag bit must then be added to each burst to
indicate when the burst is being used for signalling.

13.7 Future systems

The rapid development of mobile systems, particularly cellular networks,
has meant that mobile phones in the forms of both vehicle installations
and hand-held units are now commonplace. They have become so
popular that they have even led to a negative reaction in some quarters.
Future developments in terms of technology can be foreseen with some
confidence, although there remains the outstanding problem of portable
power supplies. However, the future development of mobile systems is
likely to be governed by political and regulatory decisions as much as by
technological and commercial considerations.

The current liberalization of telecommunications in many countries
has opened up some interesting possibilities. For example, it is difficult
to envisage several competing telecommunications operating companies
all digging up streets to lay their cables between the user terminals and a
local switching point. At most two or possibly three operators might run
competing fixed networks in the business areas of large cities. However,
if radio links are used to provide local distribution, the field is wide
open. Once the base stations have been set up, new operating companies
could start to compete with established operators, even if initially they
had only a few subscribers. Of course how many operators would
survive in such a competitive market is a matter for commercial
speculation. The obvious way to exploit such radio services would be to
offer combined fixed and mobile service; this possibility is being
examined carefully by operating companies that see a potential
opportunity to compete further with the established PTTs.

The concept of a combined fixed and mobile network is seen as an
interesting possibility, not just in providing increased competition in
telecommunications services. At present, a substantial number of people

have three or more telecommunications 'numbers', namely office, home and mobile, plus additional numbers for fax and other services. With stored program-controlled exchanges in the network, it is technically feasible to use one personal number and to route calls to the subscriber or to diversion terminals as required. However, the regulatory problems associated with such an approach tend to be extremely complex, particularly if competing networks attempt to gain the maximum commercial advantage from the interconnections with other networks.

On the other hand, it may be argued that radio links for fixed local distribution and mobile services have many disadvantages. It is doubtful if a radio link can even provide the excellent transmission that is possible with a fixed connection, and radio terminal equipment is potentially more expensive than that required for fixed connections. Furthermore, interception of radio communications is relatively easy, whereas for a fixed connection some physical and potentially detectable infringement is necessary. Above all, there remains the limitation of available radio spectrum.

Although digital techniques may compete with analogue for some time to come, they appear to have more potential for further development. As for fixed networks, the digital approach provides a basis for a network that can carry virtually all present and future telecommunications services. At present, both analogue and digital systems appear to be about equal in the efficient use of the radio spectrum although there is considerable debate about the conditions for valid comparison. In speech coding, the GSM system mentioned earlier uses coding at 13 kbit/s, but it is significant that the system definition allows for half-rate channels in the expectation that coding techniques will improve in the future to allow the volume of traffic to be doubled while retaining the same bandwidth. The basic approach is to remove as much as possible of the redundancy present in the input signals and then to introduce redundancy in the most effective manner to overcome the deficiencies of the radio path. If the system is oriented towards digital transmission, corresponding techniques may be used for other services in addition to speech. A further advantage of a digital system is that encryption may be introduced relatively easily both to provide privacy and to guard against fraudulent use of the network.

If the use of radio for combined local distribution and mobile networks is to become widespread, more radio spectrum will be required. As technology advances, new frequency bands will be brought into use but mobile systems will have to compete with other services for a share of the spectral 'cake'. The absorption bands starting at 60 GHz offer possibly the greatest opportunity for mobile and related systems. The high attenuation at such frequencies makes them useless for all but very short distance links, but this is an advantage in a mobile system based on very small cells (less than 1 km radius), because of the frequency reuse. However, the infrastructure for a system based on very small cells would be very expensive to establish. The introduction of such systems is likely to depend more on political and commercial decisions than on progress in technology.

References

Atal, B. S. (1982) 'Predictive coding of speech at low bit rates', *IEEE Trans. on Communications*, COM-30, 600–14.

Cattermole, K. W. (1986) *Mathematical Foundations for Communication Engineering*, vol. 2, p. 30, Pentech Press, London.

CCIR (1986a) 'Attenuation by atmospheric gases', CCIR Report 719–2 in *Recommendations and Reports of the CCIR*, vol. V, ITU, Geneva.

CCIR (1986b) 'Worldwide minimum external noise levels', CCIR, vol. 1, no. 670, ITU, Geneva.

Comroe, R. A. and Costello, D. J. (1984) 'ARQ schemes for data transmission in mobile radio systems', *IEEE Journal on Selected Areas in Communications*, vol. SAC-2, no. 4, 472–81.

Gilbert, E. N. (1960) 'Capacity of a burst noise channel', *Bell System Technical Journal*, vol. 39, no. 5, 1253–65.

Griffiths, J. (1987) *Radio Wave Propagation and Antennas: An Introduction*, Prentice-Hall, London.

ITU (1982) *ITU Radio Regulations*, Article 8.

ITU (1987) 'Final acts of the World Administrative Radio conference for the mobile services' (MOB-87).

Jayant, N. S. (1986) 'Coding speech at low bit rates', *IEEE Spectrum*, vol. 23, no. 8, 58–63.

MacDonald, B. H. (1979) 'The cellular concept', *Bell System Technical Journal*, vol. 58, no. 1, 15–41.

Matthews, P. A. (1965) *Radio Wave Propagation, VHF and Above*, Chapman & Hall, London.

McGeehan, J. D. and Bateman, A. J. (1984) 'Phase-locked transparent tone-in-band: a new spectrum configuration for transmission of data over SSB mobile radio networks', *IEEE Transactions on Communications*, vol. 32, no. 1, 81–87.

Pasupathy, S. (1979) 'Minimum shift keying: a spectrally efficient modulation', *IEEE Communications Magazine*, vol. 17, no. 4, 14–22.

Rice, S. O. (1948) 'Properties of a sine wave plus random noise', *Bell System Technical Journal*, vol. 27, no. 1, 109–57.

14 Broadband ISDN and future services

DON PEARSON

14.1 Introduction

This chapter is concerned with present and future teleservices which require large transmission bit rates. The CCITT proposes to handle these services via a broadband ISDN (B-ISDN) network (Griffiths, 1990; Kessler, 1990; Stallings, 1992); certain characteristics of this network have been specified by study group XVIII and others are still under study. We shall look at the way in which the thinking about this network is evolving and consider the opportunities for possible new services. At this early stage in the evolution of the network, the emphasis will be less on the detailed structural characteristics of the network and more on the potential demand for broadband services. Without a broadband network there can be no broadband services, but the provision of B-ISDN will not necessarily mean that customers will want to use it. The communication needs and behaviour of users are likely to be the determining factors that shape the network.

The chapter is structured as follows: in Sec. 14.2 there is an outline of the technical terms and philosophy of B-ISDN as envisaged by the CCITT; in Sec. 14.3 current and future envisaged broadband services are described (in particular, video); in Sec. 14.4 there is an explanation of why ATM is likely to be a useful provision; and in Sec. 14.5 there is a discussion of why some broadband services fail and others succeed.

14.2 B-ISDN

The integrated services digital network (Griffiths, 1990; Kessler, 1990; Stallings, 1992) has evolved from an earlier telephony network, the integrated digital network (IDN). ISDN extends digital transmission to the customer's premises and supports a wide range of voice and nonvoice services, via a limited set of multipurpose customer interfaces. The CCITT published recommendations for the ISDN in 1984 and several telecommunications authorities now offer it. However, the rate of penetration of ISDN has not been as rapid as was predicted in some early forecasts.

ISDN can be divided into narrowband and broadband ISDN. *Broadband* is defined by the CCITT as being capable of supporting transmission rates greater than the primary rate, which is 2.048 Mbit/s in Europe and 1.544 Mbit/s in the USA. So the consideration of B-ISDN in this chapter involves teleservices with bit rates in excess of around 2 Mbit/s. In what follows a brief introduction is given to the technical terms and philosophy of B-ISDN, as it is currently viewed by the CCITT.

14.2.1 Technical terms

Some of the important terms used to describe B-ISDN are given below, as defined in CCITT recommendation I.113 (CCITT, 1992). We shall refer to several of them later in the text. The first three definitions are concerned with the characterization of the bit rate of a teleservice as being either variable or constant:

- *Service bit rate*   The bit rate which is available to a user for the transfer of user information.
- *Variable bit rate (VBR) service*   A type of telecommunication service characterized by a service bit rate specified by statistically expressed parameters which allows the bit rate to vary within defined limits.
- *Constant bit rate (CBR) service*   A type of telecommunication service characterized by a service bit rate specified by a constant value.

The next set of definitions provides a means of classifying B-ISDN services into two main types, depending on whether the information flow is unidirectional or bidirectional.

- *Distribution service*   A service characterized by the unidirectional flow of information from a given point in the network to other (multiple) locations. Distribution services are subdivided into two classes: distribution services without user individual presentation control and distribution services with user individual presentation control.
- *Interactive service*   A service which provides the means for bidirectional exchange of information between users or between users and hosts. Interactive services are subdivided into three classes of services: conversational services, messaging services and retrieval services.

The third group of definitions relates to a new way in which information can be transferred over a B-ISDN network:

- *Transfer mode*   Aspects covering transmission, multiplexing and switching in a telecommunications network.
- *Block*   A unit of information consisting of a header and an information field.
- *Cell*   A block of fixed length. It is identified by a label at the asynchronous transfer mode layer of the B-ISDN protocol reference model.
- *Asynchronous transfer mode (ATM)*   A transfer mode in which the information is organized into cells; it is asynchronous in the sense that the recurrence of cells containing information from an individual user is not necessarily periodic.

The last definition, of ATM, introduced a significant innovation in telecommunications network transport; it has its origins in computer communication networks (Waters, 1991), where bits tend to be generated in bursts and are therefore best transported in packet form. A cell is really a packet of fixed length. ATM provides a means for variable bit-rate transport not only of computer data but also of speech and video signals. The phrase 'not necessarily periodic' implies that either VBR or CBR transmission can be used.

14.2.2 The philosophy of B-ISDN

The following selected extracts from CCITT recommendation I.121 (CCITT, 1990) deal with the aims of B-ISDN:

- The main feature of the ISDN concept is the support of a wide range of audio, video and data applications in the same network. A key element of service integration for an ISDN is the provision of a wide range of services to a broad variety of users utilizing a limited set of connection types and multipurpose user/network interfaces.
- In the context of this recommendation, the term B-ISDN is used for convenience in order to refer to and emphasize the broadband aspects of ISDN. The intent, however, is that there be one comprehensive notion of an ISDN which provides broadband and other ISDN services.
- Asynchronous transfer mode (ATM) is the transfer mode for implementing B-ISDN and is independent of the means of transport at the physical layer.
- The underlying ATM of the B-ISDN provides some specific, advantageous facilities:
  — high flexibility of network access due to the cell transport concept and specific cell transfer principles,
  — dynamic bandwidth allocation on demand with a fine degree of granularity,

— flexible bearer capability allocation and easy provision of semi-permanent connections due to the virtual path concept,
— independence of the means of transport at the physical layer.
• The deployment of B-ISDN may require a period of time extending over one or more decades, as operators seek to find the most economic means of evolving to the B-ISDN. These evolutionary phases (e.g., deployment of metropolitan area networks, passive optical networks, local area networks and also satellite based networks) will need to be harmonized with the overall B-ISDN concepts ensuring the continued support of existing interfaces and services and be eventually integrated with the B-ISDN. In these evolutionary phases appropriate arrangements must be developed for the interworking of services on B-ISDN and services on other networks.

The CCITT has established user–network interfaces (UNIs) at 155.520 Mbit/s and 622.080 Mbit/s (CCITT, I.413 1992; I.432 1990), with arrangements for packing the ATM cell structure into a synchronous digital hierarchy (SDH) frame.

14.2.3 B-ISDN services

At present there is no agreed view as to which broadband services are likely to come into being or which are likely to become dominant. The CCITT has, however, made a list of possible services in I.211 (CCITT, 1992), a summary of which follows. The initial classification is into interactive and distribution services; each is then further subdivided.

Interactive services

Conversational services

• Broadband videotelephony
• Broadband videoconference
• Video surveillance
• Video/audio information transmission services
• Multiple sound programme signals
• High-speed unrestricted digital information transmission service
• High-volume file transfer service
• High-speed teleaction (control, telemetry, alarms)
• High-speed telefax
• High-resolution image communication service
• Document communication service

Messaging services

- Video mail service
- Document mail service

Retrieval services

- Broadband videotex
- Video retrieval service
- High-resolution image retrieval service
- Document retrieval service
- Data retrieval service

Distribution services

Distribution services without user individual presentation control

- Existing-quality television distribution service (PAL, SECAM, and NTSC)
- Extended-quality television distribution service
- High-definition television distribution service
- Pay television
- Document distribution service
- High-speed unrestricted digital information service
- Video information distribution service

Distribution services with user individual presentation control

- Full-channel broadcast videography (text, graphics, sound, still images)

As can be seen, approximately three-quarters of the items in the list involve some form of image transmission; this reflects a widespread view that images and video signals will play an important if not dominant role in future broadband services.

14.3 Network requirements: video services

Because of their impact on the future growth of B-ISDN, it is of considerable interest to know which of the possible video services listed by the CCITT are likely to expand and to make major demands on the network. In this regard it is appropriate to take a realistic look at the past successes and failures of video as a broadband communication service. It is of course neither possible nor desirable wholly to separate visual from auditory or tactile means of communication; all form part of the complex multimedia panoply of physical ways in which humans

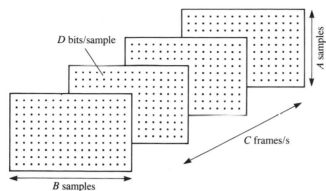

Figure 14.1 Influence of scanning parameters on the bit rate of a visual service. For noninterlaced video this rate is approximately equal to the product *ABCD* and for line-interlaced video *ABCD*/2. Both rates are affected by factors such as the duration of blanking intervals

communicate. Vision, however, tends to need significantly more transmission capacity than the other sensory modalities because of its multidimensionality. Moving images have two dimensions of space and one of time; when scanned, the output from an individual picture element (pixel) is effectively multiplexed with the outputs of thousands of other pixels. This multiplexing leads to very high bit rates for high-resolution images. The network capacity required thus depends on the scanning parameters of the system, which we examine first in a mechanistic way and later in relation to user needs.

14.3.1 Bit rate as a function of image scanning parameters

There are four fundamental parameters, *A*, *B*, *C*, and *D*, which influence the bit rate of a video signal, as illustrated in Fig. 14.1. *A* and *B* together determine the size and sharpness of the window which the telecommunications channel opens up on the world; *C* determines the temporal and *D* the grey-scale resolution of the image. The overall raw bit rate (before any data compression) is proportional to the product *ABCD* of these four parameters.

Table 14.1 is intended to give an idea of the wide range of values of the *ABCD* parameters encountered in current and future envisaged visual telecommunication services. A particularly wide variation is found in the parameters *A* and *B*; for videotelephony a 64 × 64 low-resolution image of the face may suffice, but for high-definition TV (HDTV) (*Image Communication*, 1990) and super high-definition TV (SHDTV)

Table 14.1 Approximate values of the parameters A–D with resulting uncompressed bit rates for different visual services[1]

System	A	B	C	D	ABCD (Mbit/s)	ABCD (kbit/s)
Low-quality videophone	64	64	8	6	0.2	
Common intermediate format (CIF) videoconference[2]	352	288	30	8	24	
Digital broadcast TV	720	576	25	8	83	
HDTV	1920	1150	50	8	883	
SHDTV[3]	2000– 10 000	2000– 10 000	50	8	1600– 40 000	
Facsimile	1200	800	0.01	1		9.6
	1200	800	0.07	1		64
	1200	800	2.1	1		1984

[1] The parameters given are for the luminance signal only; the addition of colour in the form of chrominance signals increases the rate further. For comparison, all services are shown with noninterlaced scanning; however, it is well known that in broadcast TV, for example, interlacing is used (interlacing is here regarded as a data-compression technique).
[2] An image size of one-quarter CIF (176 × 144 pixels) is used in lower resolution systems. Also the chrominance components for a CIF image are typically sampled at one-quarter CIF.
[3] New development under study.

(Ohta and Ono, 1991), which are intended to provide wide-angle, high-resolution windows, A and B may both be 1000–2000 samples or more. This results in a raw bit rate measured in Gbit/s.

With these very large bit rates, a variety of image data-compression methods are currently under investigation. Among them are predictive coding, conditional replenishment, transform and wavelet coding, vector quantization, subband coding, and model-based coding (Pearson, 1991). Figure 14.2 illustrates the wide range of bit rates generated by video signals. In the figure both colour and data compression have been taken into account; the range of bit rates associated with each class of service is a consequence of different scanning standards and data-compression methods. The lower bit-rate limits of each service are not meant to be definitive; they are constantly being revised downwards as data-compression techniques improve. Recently there has been a move, particularly in the USA, to all-digital transmission of HDTV, in which compression ratios of the order of 100:1 are being considered in order to utilize existing analogue broadcast TV channels (*Image Communication*, 1992).

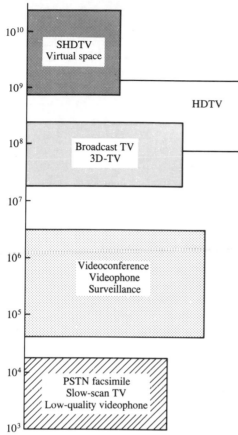

Figure 14.2   Bit rates (including data compression) for video and multimedia services

14.4 Video bit-rate variation and ATM transmission

14.4.1 ATM video

When high-compression coding techniques are used with video signals, a variable bit rate tends to be generated, owing to the interaction between the scene content and the data-reduction process (Hughes, 1983; Ghanbari and Pearson, 1989). A typical form of a modern high-compression video coder is illustrated in Fig. 14.3. It uses a concatenation or hybrid of several techniques. The first is interframe prediction, which may be motion compensated; this results in a nonnegligible prediction error signal only in those parts of the image where there is unpredictable movement. The prediction error signal is

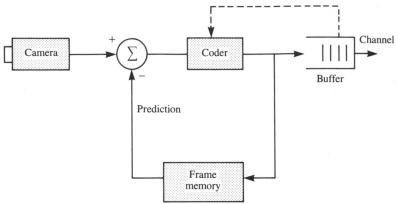

Figure 14.3 High-compression video coding using interframe prediction. The coded output is variable as a function of time and is therefore buffered for transmission over a CBR channel

therefore a highly time-variant signal; it is coded using a spatial or intraframe method such as transform coding, followed by variable word-length coding. Because the resulting bit stream is variable, it needs to be buffered for transmission over a CBR channel, as illustrated in Fig. 14.3. When the buffer fills up, a feedback path signal coarsens the quantization in the coder, causing a reduction in the picture quality. This method of coding forms the basis for CCITT recommendation H.261 (CCITT, 1990).

As noted earlier, one of the attractive features of B-ISDN is the provision for ATM transmission. In this cell-based form of transmission, bits can be transmitted between source and destination at a variable rate. If the network can handle the peaks in the video bit rate, the final buffer stage in Fig. 14.3 can be eliminated, leading to the implementation shown in Fig. 14.4. Here the coded video is packetized into fixed-length ATM cells, which have a 5-byte header and 48-byte information field. Because it takes a finite time to fill up the 384-bit capacity of a cell before it is transmitted, there is still a small buffering action; however, this is not sufficient to absorb the large components of bit-rate variation Ghanbari and Pearson, 1989). This is achieved in the network itself through statistical multiplexing (as discussed in Chapter 6) of many VBR sources.

Among the potential advantages of VBR transmission of video signals using ATM are the following:

- Because buffering is reduced or eliminated, lower values of end-to-end transmission delay can be achieved.
- If the network can absorb the variations in bit rate, it should be possible to maintain a constant picture quality during the time

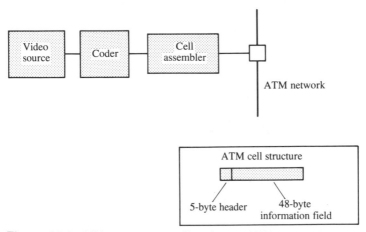

Figure 14.4   Video source coding for an ATM network

period of the transmission, even though high data compression is
used.
- If an appropriate level of statistical multiplexing can be achieved,
  VBR transmission may be more efficient than CBR transmission
  because more sources can access the network at the same time.

14.4.2 ATM network utilization

The last point in the previous section, namely the efficiency of statistical
multiplexing, is an important if not a crucial one for the success of the
ATM concept which has been designed into B-ISDN. To consider this
further therefore, suppose that N independent VBR video sources, each
of mean bit rate M bit/s, access a network of total capacity Q bit/s (Fig.
14.5). This is a simplification of the true multiservice situation in which
voice, data, and video signals share a B-ISDN network.

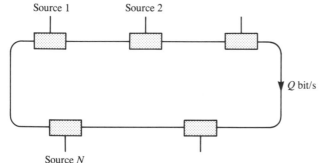

Figure 14.5   Simultaneous use of a B-ISDN network, of total capacity
Q bit/s by N variable bit-rate video sources, each of average rate M
bit/s

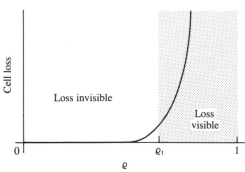

Figure 14.6 Cell loss in ATM video networks as a function of the network loading, expressed in terms of the utilization factor ρ. Below a critical value ρ_t the cell losses are invisible

The network utilization (factor), ρ, is given by the expression:

$$\rho = NM/Q \qquad (14.1)$$

ρ provides a measure of how much the network is loaded. Suppose N is large, so that ensemble averaging is effective. Then if ρ is gradually increased from 0 to 1, there is an increasing risk that bit-rate peaks in individual video sources may coincide to drive the total required instantaneous bit rate for all sources above the network capacity Q. If this happens, the network protocols will be forced to queue cells at the network nodes, introducing delay. It has been found (Eade *et al.*, 1991) that the characteristic of multiplexed ATM networks with increasing ρ is as shown in Fig. 14.6. Below a certain threshold value ρ_t (which is estimated to be about 0.7 for a large number of sources) the delays result in an invisible discard rate of cells. Above ρ_t there is a visible discard rate, leading to a loss of picture quality. This loss is more rapid when there is a movement in the scene; for scenes with considerable movement the visibility of the impairment can increase from 'just perceptible' to 'annoying' as ρ increases from 0.7 to 0.8 (Hughes *et al.*, 1993).

The statistical multiplexing gain G of VBR over CBR transmission is calculable if the peak/mean bit-rate ratio v of the VBR video signal is known. For broadcast signals using current compression techniques the ratio of the peak (averaged over a frame) to the long-term mean has been measured as about 4.7, hence:

$$\begin{aligned} G &= v\rho \\ &\approx 4.7 \times 0.7 = 3.3 \end{aligned} \qquad (14.2)$$

The significance of $G = 3.3$ is that 3.3 times as many video signals can access a B-ISDN network in VBR mode than in CBR mode. This assumes, however, that the total number of video sources is large.

Future coding techniques such as model-based coding are likely to make coded video even more bursty, increasing the peak-to-mean ratio and statistical multiplexing gain (Pearson, 1992).

14.4.3 Two-layer coding

A further development in ATM video is in the use of two-layer coding, in which the coded bits are split into two cell streams, one having guaranteed transport within a certain time period (Ghanbari, 1989; Morrison, 1990). The more structurally important image information is contained in these guaranteed cells; the network protocols then discard the low-priority cells at times of overload, based on detection of a cell loss priority (CLP) indicator in each transmitted cell.

14.4.4 CCITT Experts Group on ATM video

In July 1990, a CCITT Experts Group for ATM video coding was established to consider standards. The group was set up by study group XV, 'to investigate new possibilities for video coding offered by service support on the B-ISDN, and to develop appropriate coding algorithms'. It aims to make a final recommendation, provisionally known as H.26X, by 1994.

14.5 The demand for broadband video services

14.5.1 Categorization of performance

We turn now to the question: what factors make for a successful broadband service? As a first step in answering this, let us attempt to categorize the video and multimedia services in Fig. 14.2 on a relative success scale:

- *Failures* Videotelephone
- *Marginal successes* Videoconference, videotex
- *'Sleeper' now successful* Facsimile
- *Highly successful* Broadcast TV
- *Future unknown* HDTV
- *Being developed* SHDTV, 3D-TV, Virtual space/reality systems

14.5.2 The analysis of success and failure

The attempt to introduce a person-to-person videotelephone service in the late 1960s in the USA (with other countries following suit) resulted in failure in the early 1970s, though a number of technical advances in image coding were pioneered at the time. The service offering was subsequently withdrawn. Subsequently it was reasoned that the high service costs of two-way vision could more easily be justified if groups rather than individuals were connected, since the travel savings were many times greater. Videoconference services have indeed been successful, but not hugely so; large numbers of people still like to travel, though during the 1991 Gulf war it was reported that greater use was made of videoconferencing. CCITT recommendation H.261 on video coding presented manufacturers with an international opportunity to market videoconference services; although some interest has been shown, there is no evidence to date that a breakthrough has been made or a take-off point reached.

Facsimile (fax) is an interesting case study in visual telecommunication, though it is as yet largely narrowband. It was a Cinderella service until the 1980s, when it assumed a new importance and grew rapidly. Earlier terminals required the document to be wrapped round a drum for scanning; nowadays terminals are no more difficult to use than a photocopier. The time and cost of sending a page have improved considerably through the use of data-compression methods, and standardization has allowed easy international transmission. Fax is thus a visual teleservice that has broken through a threshold in the demand curve and is now established.

Broadcast television must count as the biggest success story in broadband telecommunication. There are very few countries left in the world without this service, and its attractions are well known and documented. As a model of a successful multimedia service, it invites a thorough understanding of both its technical and its psychological factors. Space does not permit a detailed consideration, but perhaps a few salient points could be made:

- Viewed at the CCIR-recommended distance of six times picture height, broadcast television presents a $12° \times 9°$ window with resolution equal to about half that which the eye is capable of resolving.
- Through this window a very wide range of visually interesting subject matter can be seen (in 2-D and colour), including sport, news, and drama. Human beings have always been interested in watching such events and have in the past been prepared to pay money to do so, either live or on film.
- The cost of viewing (receiver amortization plus a UK television licence) is of the order of 2p/hour of broadcasting; this compares very

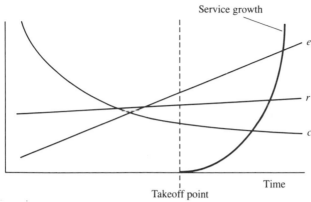

Figure 14.7 Assumed relationships between service growth, ease of use, *e*, realism, *r*, and cost, *c*

favourably with the cost of a newspaper at, say, 40p, which might be read in about an hour. In countries such as the USA, where there is no television licence fee, the comparison is even more favourable.

14.5.3 Factors influencing the growth of new services

From a study of these cases, a growth relationship (Fig 14.7) can tentatively be established for a broadband visual or multimedia service. Firstly, the assumption is made that with the passage of time there is a decrease in the cost, *c*, of providing the service, this cost being relative to average income. With increases in real income, improvements in transmission efficiency through the use of data-compression techniques, and reductions in hardware costs, this is a fairly realistic assumption.

Secondly, it is assumed that the service becomes easier to use with time through improvements in such items as the human interface; this is shown by the steady increase in the slope of the line marked *e* in Fig. 14.7. As already mentioned, this kind of improvement has been obtained in services such as facsimile; it is also true of other services such as broadcast television, for example through the use of automatic tuning, or remote-control devices that allow the user to change channels and control audio-visual parameters.

The third factor *r* (realism) is perhaps the most interesting. We suppose that the telecommunication system attempts to recreate the visual, sound, and tactile fields which the user would experience if located at the transmitting terminal (Fig. 14.8). If there is no perceptible difference between reality and the teleconstructed sensory fields, the realism factor might be said to be unity. Alexander Graham Bell's first distorted telephone transmissions and John Logie Baird's flickering images would rate a low but nonzero score; HDTV and SHDTV with

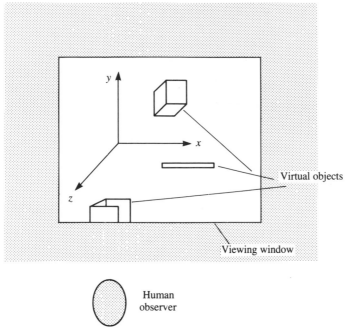

Figure 14.8   Multimedia communication systems can be evaluated in terms of their ability to recreate a virtual space, *xyz*, of sound, light, and possibly touch at the receiving terminal

stereo sound would rate much higher; while a high-resolution virtual reality system (Xu *et al.*, 1989) might rate near 1. This said, it is of course no easy matter to associate precise and unambiguous numerical scores in the range $0 \leqslant r \leqslant 1$ to any multimedia service; this requires sensory trade-offs as between sound, vision, etc., which may be context dependent and difficult to make. These trade-offs are, however, at the root of new service provision and such difficulties need to be faced.

Finally, we assume that the quantity:

$$P = nre/c \qquad (14.3)$$

represents the potential for success at any point in time, Here n is a factor indicating the need customers have for such a service, i.e., the desire to see or hear or 'touch' (in a virtual-reality space) the objects which a broadband service makes possible. Although broadcast television, as noted, is very successful, it would not be so if the entertainment and news being offered were of no interest to customers. When P reaches a threshold value, it may trigger a take-off in the growth of a new service, as illustrated in Fig. 14.7. It is not claimed that the four factors, n, r, e and c, are the only ones influencing this take-off, but that they are significant ones for engineers involved in new service development.

The failure of the videotelephone launch in the early 1970s can be considered in terms of Eq. (14.3) and Figs. 14.7 and 14.8. The cost of the service was relatively high, the ease of use was restricted by the lack of an international standard, and the realism was monochrome 2-D with an approximate 250×250 resolution in a viewing window of about $8° \times 8°$. The most difficult and controversial point in videophone assessment, however, is the need factor, n. A view has prevailed for many years among certain groups of communications specialists that people have a desire to look at one another while talking on the telephone, indeed some of the early pioneers of television in the USA thought that its growth would come as an adjunct to the telephone. Studies by social psychologists of the way in which human beings look at one another (Argyle and Cook, 1976) are not as well known among systems designers as they should be; they reveal many interesting facets, among them that we look at people's faces less and less, the closer we are to them. There are also national, status- and time-dependent factors in human gaze. It is not clear which of the factors n, r, e or c, was crucial in the initial failure of the videotelephone, but n is probably lower than was thought at the time. In recent years there has been a good deal of further technical development in the videotelephone and videoconferencing areas, for example in data compression and standardization. Much of this is predicated on the assumption that n is inherently high enough such that if r and e can be increased and c diminished, the service should take off.

14.5.4 Future broadband services

There is currently a great deal of discussion of systems for bringing HDTV into the home (*Image Communication*, 1990; 1992). Regarded in the terms of Fig. 14.8, HDTV offers a wider viewing window and/or improved resolution within this window by the approximate doubling of both A and B scanning parameters, as compared with present-day broadcast television. As a broadcast service, the value of the need factor, n, for HDTV should be high; the realism, r, is significantly improved, particularly for sports and outdoor scenes. Crucial therefore is whether the ease of use, e, and cost, c, can be controlled. Currently, the receivers are very bulky, and wall-mounted displays have not yet reached an appropriate stage of development. All manner of ingenious coding and transmission schemes have been put forward to get HDTV into the home before the advent of optical fibres; many of these have, however, lowered the resolution within the viewing window.

High-quality stereo sound is now widespread and 3-D TV is under serious study (*Image Communication*, 1991). With all the attention being given to HDTV, 3-D TV has been somewhat neglected. It offers a considerably heightened sense of realism, r, at very little extra cost, c. The main problem holding back its development is the display; the ease of use, e, is low with most current types, involving as they do the use of glasses or restricted-movement lenticular displays. Virtual-reality systems provide an even greater realism, but the head-mounted displays have an even lower value of e.

Super HDTV is a fairly recent development, in which the parameters A and B are approximately doubled again over that of HDTV (Ohta and Ono, 1991). The resulting four-fold increase in image pixels pushes the raw bit rate up to around 7 Gbit/s, with data-compression methods being capable of reducing this figure by 10:1 or more. With reference to Fig. 14.8, the main improvement is that the window size and/or resolution are increased over those of HDTV. At the time of writing, 2000 × 2000 SHDTV displays exist, but image capture methods are available for still images only; it is envisaged that moving SHDTV sequences will be generated in the near future. With the difficulty experienced in getting HDTV into the home, SHDTV might at first be used in medical, advertising, and printing applications; however, in the longer term it could also be the basis of B-ISDN distribution or retrieval services of startling realism.

B-ISDN is in concept a very flexible service with a number of attractive features such as ATM. The built-in flexibility and universality of the network may however have a cost implication, namely that B-ISDN transmission could be relatively expensive compared with the less sophisticated radio, satellite, and cable networks which distribute broadband entertainment today. The slow growth in ISDN is a warning that B-ISDN costs need to be carefully considered and that efficient coding techniques should be developed which fully exploit its VBR capability. The likely balance between conversational and distribution services on B-ISDN requires careful thought in view of previous failures of video conversational services and the complex human factors involved. If future broadband video services are mainly of the distribution type, it may be possible to accommodate them on extensions to the current cable network. Competition from cable providers may therefore influence the architectural developments in B-ISDN towards simpler and less expensive routeing.

14.6 Summary

This chapter has outlined current thinking about the structure of B-ISDN and has discussed future possible teleservices on this network. It is thought likely that many of these will be video services or multimedia services in which video has a dominant role. The way in which the realism of a video service is related to the bit rate has been described; also how modern coding methods tend to produce a variable rate suitable for ATM transmission. Exerimental evidence suggests that ATM VBR transmission can be more efficient than CBR transmission if the number of network users is high.

In assessing the potential for growth of future broadband services on B-ISDN, the concept of virtual space had been used as a means for comparing reproduced sound and vision fields to those that exist at the transmission source. An equation, $P = nre/c$, has been introduced as a guide to the understanding and analysis of broadband services; this represents the potential, P, for service growth as a function of user need, n, realism, r, ease of use, e, and cost, c. The failure of some past services has been explained in terms of this model. New services such as HDTV, 3D-TV and SHDTV have also been analysed; if terminal and transmission costs can be kept low and the systems made easy to use, they appear to have a bright future. There is some reason to believe that distribution rather than conversational video teleservices will be dominant in future broadband networks; if so, the network architectures will need to evolve to take this into account.

References and further reading

References

Argyle, M. and Cook, M. (1976) *Gaze and Mutual Gaze*, Cambridge University Press.

CCITT I.113, 'Vocabulary of terms for broadband aspects of ISDN', 1992.

CCITT I.121, 'Broadband aspects of ISDN ', 1992.

CCITT I.413, 'B-ISDN user–network interface', 1990.

CCITT Draft Recommendation 'I.432, B-ISDN user–network interface— physical layer specification', 1990.

CCITT I.211, 'B-ISDN service aspects', 1992.

CCITT H.261, 'Video coding for audiovisual services at p × 64 kbit/s', 1990.

Eade, J. P., Hughes, C. J. and Xiong, J. (1991) 'Modelling multiplexed video VBR networks', in *Proc. Fourth International Workshop on Packet Video*, pp. F3-1–F3-5, Kyoto, Japan, August 1991.

Ghanbari, M. (1989) 'Two-layer coding of video signals for VBR networks', *IEEE Journal on Selected Areas in Communications*, vol. 7, no. 5, 771–81.

Ghanbari, M. and Pearson, D. E. (1989) 'Components of bit-rate variation in videoconference signals', *Electronics Letters*, vol. 25, no. 4, 285–6.

Griffiths, J. M. (1990) *ISDN Explained*, John Wiley, New York.

Hughes, C. J. (1983) 'Evolution of switched telecommunication networks', *ICL Technical Journal*, 313–29.

Hughes, C. J., Ghanbari, M., Pearson, D. E., Seferedis, V. and Xiong, J. (1993) 'Modelliing and subjective assessment of cell discard in ATM video', *IEEE Trans. on Image Processing* vol. 2, no. 2, 212–222.

Image Communication, special issue on HDTV, vol. 2, no. 3, October 1990.

Image Communication, special issue on 3D-TV, vol. 4, no. 1, November 1991.

Image Communication, special issue on all-digital HDTV, vol. 4, nos. 4–5, August 1992.

Kessler, G. C. (1990) *ISDN*, McGraw-Hill, New York.

Morrison, D. G. (1990) 'Variable bit rate video coding for asynchronous transfer mode networks', *British Telecom Technical Journal*, vol. 8, no. 3, 70–80.

Ohta, N. and Ono, S. (1991) 'Super high definition image communication—applications and technologies', in *Signal Processing of HDTV, 3, Proc. Fourth International Workshop on HDTV and beyond*, Turin, Italy.

Pearson, D. E. (ed.) (1991) *Image Processing*, McGraw-Hill, London.

Pearson, D. E. (1992) 'The implications of current video coding research for broadband ISDN', in *Proc. IEEE Workshop on Visual Signal Processing and Communications*, Raleigh, NC, 2–3 September 1992.

Stallings, W. (1992) *ISDN and Broadband ISDN*, Macmillan, New York.

Waters, G. (1991) *Computer Communication Networks*, McGraw-Hill, London.

Xu, G., Agawa, H., Nagashima, Y. and Kobayashi, Y. (1989) 'A stereo-based approach to face modelling for the ATR virtual space conferencing system', in *Proc. SPIE Conference on Visual Communication and Image Processing*, vol. 1, 365–79, Philadelphia, 8–10 November 1989.

Further reading

IEEE (1989) 'Packet Speech and Video', *IEEE Journal on Selected Areas in Communications*, vol. 7, no. 5.

IEEE (1990) *Proc. IEE International Conference on Integrated Broadband Services and Networks*, 15–18 October 1990, IEE Conference Publication no. 329.

British Telecommunications (1990) 'Audiovisual Telecommunications', *British Telecom Journal*, special issue, vol. 8, no. 3.

15 From copper to glass

PETER COCHRANE

15.1 Introduction

The growth of civilization is critically dependent upon rapid and efficient communication at a distance for its organization, control and stability. From the runners and beacons used in ancient times, to the Chappfe telegraph, electric telegraph, radio, and today's telephone and data networks, the need has always been for faster and more effective telecommunications if man's progress was to be maintained. This is probably more evident today than at any other time, with the emerging need to combine communication, computing, and control (Palfreman and Swade, 1991). Tomorrow the list is likely to include robotics, telepresence, and variants of virtual reality at least (Engleberger, 1989). It is interesting to reflect that none of this, not even a ubiquitous telephone network, would be possible with wholly copper transmission technology. It is doubtful if there is enough copper and the associated technology, or money, available to pay for everyone to have just a telephone (Cochrane, 1984). Certainly the bandwidth limitations of copper and radio rule out any significant excursion into computing, control, telepresence, or other futuristic services. Only the timely arrival of optical fibre technology during the last 15 years has given us the means to attain the dream of telecommunications for all at a price we can all afford, with the added advantage of a near infinite bandwidth and service expansion in the future (ITU, 1991).

It is perhaps understandable that the development of copper transmission and switching technology over the past 150 years has conditioned many people in the industry to the network topology, configuration, service, and performance of the kind we currently enjoy. The reality is that the status quo will not do! Optical fibre is the means of realizing the next paradigmatic change, which may well be greater than the leap from the beacon to the electric telegraph. All we have achieved so far with optics is more of the same with better performance at a lower price (Cochrane, 1990). What we have to do is rethink the problem of systems, networks, operations, and services to encompass the advances made possible by this young technology. To put this in

Table 15.1 Key features of copper
and fibre technologies

Feature	Copper	Fibre
Bandwidth	Limited	Almost infinite
Loss	High	Very low
Linearity	Good	Very good
Crosstalk	High	Almost zero
Reliability	Poor	High
Repeater spacing	2 km	> 30 km
Cost	High	Very low

Table 15.2 Principal differences between
present-day and future networks

Feature	Today	Tomorrow
Data	Poor	Rich
Use	Point to point	Multipoint
Media	Single	Multi
Bandwidth	Limited	Almost infinite
Flexibility	Low	High
Reliability	Low	High
Utility	Low	High
Cost	High	Low
Charge for:		
Bandwidth	Yes	No
Distance	Yes	No
Time	Yes	Yes
Service	Yes	Yes

perspective it is useful to draw a limited set of comparisons between the old, new, and future technologies as indicated in Tables 15.1 and 15.2.

The purpose of this chapter is to map the development of telecommunication transmission to the present day, before extrapolating into the future, and to apply a degree of educated guesswork to the likely developments we will see up to, and into, the new millennium. To do this we briefly recollect the key epochs so far with telegraphy, analogue and digital telephony over copper pairs and coaxial cable, in relation to the recent development of optical fibre. From this base we then move into the near and far future. In doing this it is difficult not to challenge some of the established wisdoms and sacred cows of the industry—so this we do on the basis that if it can be done, then probably someone will do it, and sooner rather than later!

15.2 Telegraph systems

The emergence of rapid road and rail transport, from about 1750 onwards, necessitated the development of new forms of telecommunications, allowing the rapid transfer of information ahead of the surface vehicles. Without the ability to synchronize local times and timetables and provide a means of controlling the physical transit networks, further development would have quickly been stifled. By the early 1800s, many European countries had established far-flung empires, and extensive road, rail, and shipping networks were being constructed, while in North America the migration from the east to the west coast was about to begin. The invention and commercial realization of the electric telegraph, in about the same time frame, was both opportune and necessary. Progress was rapid and by 1835 the electric telegraph network was operating globally over 64 000 km of copper wire linking countries and continents (Barty-King, 1979). Operation was strictly point to point at first with human, and later electromechanical repeaters (relays). Key disadvantages of these systems were the need for trained operators whose operation rate with Morse code was intrinsically slow, and cable characteristics which limited the information transmission to between 10 and 40 words per minute (Nahin, 1987). On the other hand, there are those who argue that Morse code was one of the most efficient means of human communication—as a messaging system, it removed redundancy and got straight to the point! It was however very expensive. For an average man to send a message 30 km across England would have cost a week's wage at that time.

While at first sight wire telegraphy might appear crude, the reality is that it rose to a peak of sophistication during its long history. Much of the seminal work on copper cable characteristics was directed towards expanding the capability and performance of this technology (Nahin, 1987). The principles of TDM were established, including many of the synchronization and coding techniques we consider modern today. FDM was also developed to accommodate more telegraph channels on a single cable pair (Bylanski and Ingram, 1980). Also, the first evidence of crosstalk, which would become a prime limitation for telephony, and the recognition that balanced pair circuits were necessary, occurred during this era (Campbell, 1937). As a final perspective it is perhaps worth noting that the American Navy has only abandoned radio telegraphy within the past five years, whereas the British Navy plus many others still use it, as do many maritime operators and developing countries. Further, the arrival of the fax machine has only recently seen the suspension of the telegram and demise of telex services. A history spanning over 150 years—which is unlikely to be bettered by any of today's technologies!

15.3 Telephony

The invention and rapid deployment of the telephone in 1875 largely overcame the operator limitations experienced with telegraphy, and natural language (analogue) communication soon became dominant. As with telegraphy, the early telephony systems were strictly point to point and of limited range. The need for effective repeaters was soon realized as the work of Heaviside *et al.* demonstrated the fundamental limits to transmission over copper lines even with continuous and lumped loading (Nahin, 1987). To span the USA from east coast to west it was estimated that copper conductors up to 2.5 mm in diameter would be required for a single telephone line. With the emergence of suitable electromechanics (in about 1900) and electronics (in about 1915), the means of providing automatic switching and realizing repeaters for extended transmission distances became universally available. This was followed by the introduction of FDM in the 1920s and 1930s (Bylanski and Ingram, 1980).

Probably the key development beyond the introduction of electronic amplification was the realization of low crosstalk overhead wire lines and composite pair cables. Interestingly, the transposition patterns used in the overhead routes were based on Walsh functions to achieve spatial orthogonality, and hence low crosstalk coupling (IEE, 1984). For copper pair cables it was more convenient to use different sinusoidal twist rates for individual pairs and groups of pairs. This was followed by the development of coaxial cables for FDM systems, ultimately transporting up to 10 800 simultaneous analogue speech channels, far beyond (by about 50-fold) the original cable design requirement (POEEJ, 1973).

The development of PCM and transistor-based electronics jointly enabled the realization of TDM systems that could most economically combat the limitations of copper pair crosstalk and attenuation (Cattermole, 1969). By about 1960 PCM systems were becoming established, and in the 1970s they became widespread. This in turn led to the development of digital transmission systems at higher rates which had well defined and internationally agreed standards by the mid-1970s. At this time, repeater spacings had fallen to 2 km for 2.048 Mbit/s on pair cables and 140 Mbit/s on coaxial cables (Reeves, 1978). Systems operating at 565 Mbit/s rates were also being investigated, but required repeaters at 1 km intervals (Crank and Cochrane, 1980). During this same period a number of administrations had completed economic studies that established the combination of digital transmission and switching as the most economic (Flood and Cochrane, 1991). This is reflected in Fig. 15.1 which shows the options considered and their relative costs. Programme decisions made at that time are only now seeing their full realization, with the near-full digitalization of national and international networks in the developed world. (A point to be recognized here is that, historically, the fixed assets of a telephone

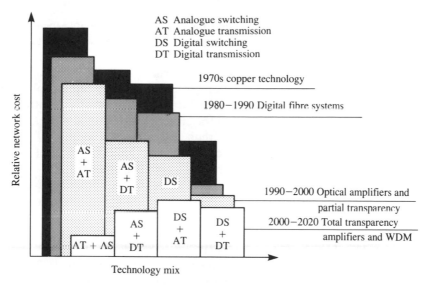

1970s   Assumes twisted copper pair PCM and coaxial digital systems at 140 Mbit/s
1980–1990   Assumes twisted copper pair and fibre PCM with fibre digital systems at 565 Mbit/s
1990–2000   Assumes a wholly fibre network with optical amplifiers and
            transparency between nodes
2000–2020   Assumes total transparency for both AT and DT with WDM and
            optical switching

Figure 15.1   Network costs with projected technology progression
relative to 1970

company (telco) are generally measured in £Bn, and any general network
changes take between 5 and 10 years to complete. Before fibre
technology, the ratio for long lines would be of the order 50 per cent
transmission and 50 per cent switching. In a modern network
transmission may now be as little as 10 per cent. Overall, about 50 per
cent of all resources can be expected to reside in the access network.)

By the early 1980s it became clear that optical fibre was the most
economic solution for the future and routes were installed on a grand
scale. In the UK for example, 2.3M km was installed to 1993—which
represented 20 per cent of the world total—carrying 85 per cent of all
UK traffic. Globally, over 55 per cent of all long-line traffic is now
transported by fibre (Cochrane *et al.*, 1991) making it the dominant
medium. The next challenge, and by far the more difficult, is the
extension into the access network with fibre to the office and home.
While the office in the medium- to large-scale site has been economically
accessible by fibre for telephony alone for some time, the same has not
been true for the small site and particularly the home, until recently.
Present constraints tend to hinge on political and/or regulatory factors
rather than technological limitations. In short, the solutions are to hand
for this last stretch in the deployment of fibre (BTTJ, 1989).

During the last 100 years the telephone has been the principal means of human telecommunication with over 850M stations now distributed across the globe (note that this distribution is very uneven in that approximately 85 per cent of telephones are with about 15 per cent of the population in the developed nations). Connection is generally point to point with switches/offices/exchanges located at demographic centres, and usually no more than 7 km from the customer site, providing a standard 3.5 kHz (64 kbit/s) speech circuit. All of this will see radical change with the introduction of fibre into the local loop. But most of all, it is in the service arena, where there is a high demand for nontelephony services, that we can expect to see the pressure for change. It is already clear that telephony is likely to be eclipsed as the primary means of communication by the year 2000. This is 10–15 years earlier than the estimates of only 5 years ago. The acceleration has been stimulated by the personal computer (PC) on every desk and the extension of computer facilities into every office, and many homes. This trend is likely to increase with the realization of new entertainment services, the use of robotics, and telepresence. Fibre in the local loop is now therefore essential for any future successful telco (ITU, 1991).

15.4 Copper in the local loop

The local loop is not only the last bastion of copper technology, it is also the area that has seen most global investment in the past. Established telcos have 50 per cent or more of their investment revenues dedicated to this part of their network. But, quite perversely, it is the portion of a network that is (traditionally) the least efficient as a revenue earner because of the low utilization and high, up-front, maintenance and operating costs. The technology is essentially very simple—a pair of twisted wires from each telephone back to a centrally located switch. The distance from telephone to switch is limited by the network signalling and transmission limits, which in turn are dictated by wire gauge, insulation, ohmic resistance, crosstalk, and impulsive and random noise (Griffiths, 1983). A global average distance is 2 km, with the peak seldom exceeding 7 km. This limitation has led to networks with thousands of dispersed switches and tree and branch cable distribution out to customer sites (Hughes, 1986). The interconnection of such large numbers of switches then dictates further switching layers organized in a hierarchical fashion with increased density of traffic and thus efficiency. This is the area where optical fibre has effectively impacted so far.

Given the size of investment in the established copper local loop, it is essential that this resource is fully exploited. However, in many countries almost every aspect of the market, regulation, competition, customer

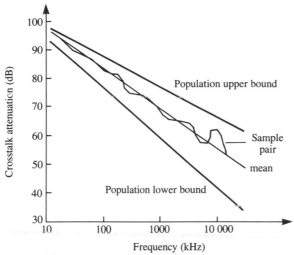

Figure 15.2 Pair cable crosstalk characteristics

demand, and new services has become clouded by government intervention (IEEE, 1989). Optical fibre for telephony alone is difficult to achieve economically (British Telecom, 1989). With a combination of telephony, wideband services, and CATV there is no problem! The situation can be neatly classified into two camps: the established telco precluded from all but the telephony market; and the newcomers who are allowed to provide all service types. The first group has by far the more difficult task and is thus still investing in extending its copper infrastructure by increasing existing capacity. Ultimately, however, it will be necessary for them to move into fibre through the *passive optical network* (PON) route as described later (Cochrane and Brain, 1988).

The established copper local loop line plant was originally designed for analogue operation with 4 kHz speech channels. Present ISDN requirements have dictated 144 kbit/s duplex operation at distances up to 4–5 km (Griffiths, 1990). Achieving such a performance improvement has necessitated major developments in signal processing technology with real-time adaptive equalization and echo cancellation techniques to overcome the variability of the copper cables (Cox and Adams, 1985). The sample copper pair characteristics given in Fig. 15.2 give an indication of the degree of variability experienced in the copper environment.

Even higher bit rate systems have been developed, with 0.5 Mbit/s and 2 Mbit/s being provided over reduced spans up to 2 km. Exceeding this level of provision is a lost cause on most installed cable and is prohibitively expensive, with the need to employ extensive testing and pair selection techniques, plus vastly increased levels of signal processing electronics. A more profitable approach involves the joint use of fibre

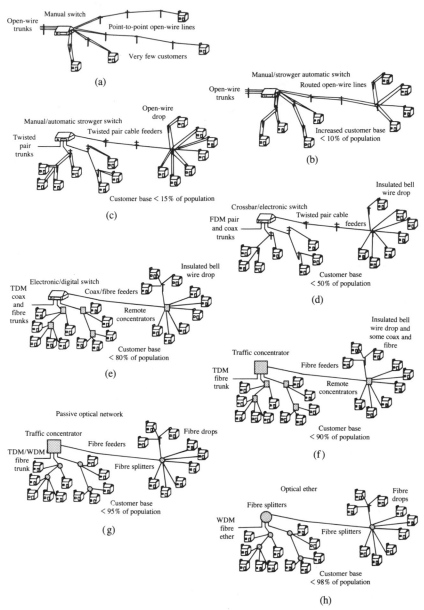

Figure 15.3   The migration of fibre into the local loop: (a) pre-1900; (b) 1900s; (c) 1920s; (d) 1950s–1960s; (e) 1970s–1980s; (f) 1980s–1990s; (g) 1990s–2000; (h) 2000–2010

and copper as indicated in Fig. 15.3 with a gradual roll out to meet the changing needs of the customer and telco (IEEE, 1991). Ultimately, this migratory path terminates in the full provision of a PON and a vast reduction in the amount of switching plant. North America is probably further down this route than any other region with the use of carrier service areas (CSAs), but other countries are also now moving in this direction with a full catalogue of tried and tested PON technology available to meet any intermediate requirements (IEEE, 1989, Cochrane and Brain, 1988).

Let us quantify the scale of the problem on a global basis in the following way:

1   There are approximately 850 million telephones with an average distance to the switching site of 2 km. This gives a total of 1.7×10^9 km of copper pair installed.

2   The global manufacturing capacity for optical fibre currently stands at 5×10^6 km/year therefore to manufacture enough fibre to replace all the copper 'one on one' would take:

$$\frac{1.7 \times 10^3 \text{ km}}{5 \times 10^6 \text{ km/year}} = 340 \text{ years}$$

and this assumes duplex operation, i.e., one fibre replaces each twisted pair! If simplex operation is used then the time to manufacture is doubled to 680 years. Even if the world manufacturing base is doubled, or increased five-fold, the time and investment implications are prohibitive.

Sharing the massive bandwidth available on fibre among many users in a PON configuration (see Fig. 15.4) realizes a much better alternative with the time to manufacture reduced by at least an order of magnitude. Moreover, it should also become apparent that merely aping the copper technology with thousands of copper pairs being replaced by thousands of optical fibres has significant reliability implications. A single fibre carrying all the traffic in a cable turns out to be more reliable in practice than multiple fibres in parallel. When a cable is damaged it takes far longer to restore thousands of fibres than it does a single fibre! This is the converse of our previous experience with copper cables since we previously never had the option of multiplexing all the traffic onto a single pair.

15.5 Signal format

The demand for nontelephony services is increasing with the dual realization of low-cost transmission capacity with optical fibres and sophisticated digital terminal equipments for the end users. In short, both customer/service demand and technology availability are providing the motivation to change the nature of telecommunications (ITU, 1991).

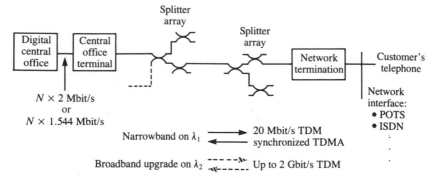

Figure 15.4a. Passive optical network (PON) principle

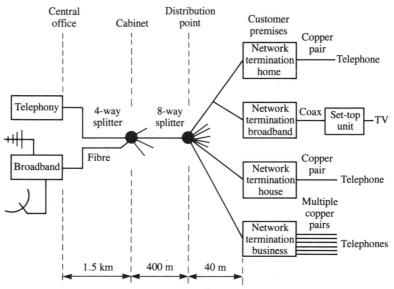

Figure 15.4b. PON architectures for telephony (TPON) and broadband (BPON) services

In the midst of this revolution it is worth while contemplating a further feature of communication history, and that is the cyclic alternation between digital and analogue forms. Before the development of natural language, the 'grunt or gesticulation' was no doubt the simplest indicator of man's agreement or displeasure (Bronowski, 1973). Man has since progressed to use the tally stick, smoke signals, drums, music, the written word, Morse code, telephony, PCM, and visual communications (ITU, 1991). In parallel, computation has also followed an alternating analogue–digital history. We might thus presuppose that this alternation will continue far into the future—and may even be encouraged by the inherent qualities of optical fibre communications and photonic

computing. There is certainly evidence that increasing complexity and
future requirements may necessitate the use of analogue techniques to
solve the problems of achieving a useful level of artificial intelligence,
associative computing, and control systems (Palfreman and Swade, 1991;
Cleick, 1987). Optical technology looks ideally placed to meet this
challenge, as previously indicated in Fig. 15.4, should it prove necessary.

15.6 The emergence of optical fibre

During the early 1970s it became clear that optical fibre would become
viable as a transmission medium, and a number of experimental systems
were constructed. Not surprisingly, optical receiver and transmitter
modules were merely attached to existing copper line repeater circuitry
to provide optical regeneration. However, aspects such as poor pulse
shaping, timing extraction, and jitter now became less critical due to the
inherent wide bandwidth and low loss of the fibre path (Brooks, 1980).

From about 1975 onwards, optical transmission started to emerge
from the feasibility stage and appeared in a number of impressive field
trials and demonstrations that clearly indicated that they were superior
to their copper predecessors. Progress was rapid, with the dual
development of multimode fibre LED technology with single-mode fibre
and lasers. Repeater spacings soon began far to exceed copper systems
and the economic viability was established. The fundamental nature of
the systems, however, remained much the same as copper with electronic
regeneration, with restricted bit rates, and direct point-to-point
operation (IEEE, 1986). Copper systems were thus merely replaced one
on one.

The prime reason for the rapid advance of fibre systems was that they
offered a dramatic cost reduction when compared with their copper and
radio predecessors, as they required far fewer repeaters and used far less
raw material and energy. Consequently, there were significant reductions
in the number of surface stations required and the amount of power
feed equipment, duct space, and maintenance. Fibre systems also became
viable at a time when copper cables and the microwave radio spectrum
were naturally approaching exhaustion. Moreover, the ultimate capacity
of the coaxial cables had also been established to be of the order of
1.2 Gbit/s for any wide-scale exploitation (Crank and Cochrane, 1980).
Coupled with the fact that the earthenware duct network was also
becoming full, some significant limits to growth were rapidly being
approached. It is interesting to note that it is the duct network and the
associated civil engineering that constitute the single largest network
investment of any network operation and thus its full exploitation is
essential.

Apart from its very small physical size, optical fibre has two key attributes that make it very nearly the ideal transmission medium: very low loss (less than 0.4 dB/km) and an almost infinite bandwidth (~ 50 000 GHz). Exploiting these directly to supersede the established copper network point-to-point links has already realized dramatic operational cost savings. However, fibre has a great deal more to offer (Cochrane *et al.*, 1991).

15.6.1 The ultimate bit rate

Optical fibre systems in service today use relatively crude intensity modulation and access less than 0.001 per cent of the inherent transmission capacity of the fibre. It has been demonstrated that pulse transmission up to 30 Gbit/s is both feasible and possible, while WDM up to 100 Gbit/s has also been demonstrated (Chidgey and Hill, 1989). So an approximately 0.2 per cent utilization has been practically demonstrated, while theoretical studies suggest a further order of magnitude may be possible. If there is demand for such capacities in the future, they would be inaccessible without the use of photonic amplifiers (IEE, 1989). High-speed electronics at rates up to 10 Gbit/s are currently possible but difficult. It may be possible to reach 20 Gbit/s, but still higher rates look extremely difficult, if not doubtful, at the time of writing.

15.7 Comparison with radio

Today's optical systems tend to use broadband LED (~ 20 000 GHz linewidth) or Fabry–Perot lasers (~ 500 GHz linewidth) sources, with APD or PIN receivers to detect binary signals. Some analogue systems have been deployed for specialized applications (CATV) but these are currently in the minority. For long-line transmission the migration towards single-mode fibre rather than the earlier favoured multimode variety is now complete, and the relatively crude 'spark transmitter' technology employed in today's systems has progressed through three broad generations operating in the 850, 1300 and 1550 nm windows (IEEE, 1986). The fourth-generation systems are likely to use narrow linewidth or coherent optics—the first step towards present-day radio signal formats and processing. The introduction of photonic amplifiers means that total transparency of the fibre path is also realizable (Cochrane and Todd, 1991). There is also an obvious and interesting synergy with radio systems now becoming apparent.

The history of radio systems has been critically dependent on the development of coherent energy sources to exploit fully the bandwidth available. This is put in perspective by Fig. 15.5 where the rate of development across the electromagnetic spectrum from 1832 to the

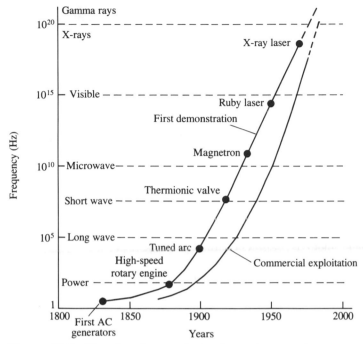

Figure 15.5 The development of coherent generators

present day can be seen. In contrast, optical systems have not exhibited the same reliance upon coherent sources as their radio and copper line predecessors, but they have still eclipsed all other media as the prime bearers for long-line transmission. This is because of the vast bandwidth of optical fibres and the containment of the guided energy. A comparison of microwaves and optics in this respect may be made by considering the transmitter power P_T and receiver power P_R for a given aperture:

$$\frac{P_R}{P_T} = \frac{\theta_R}{\theta_T} \tag{15.1}$$

where θ_R is the solid angle subtended at the receiver
   θ_T is the solid angle subtended at the transmitter

For a microwave system at X-band with a 3 m dish and 0.5° beamwidth, compared with a 1300 nm optical system with a 0.0005° beamwidth, the ratio of powers is then about 10^{-6}. Optical fibre systems thus require only about 10^{-6} of the power of a microwave system through the concentration and guidance of the energy transmitted.

A further interesting comparison may also be drawn from the ratio of carrier frequency to linewidth (i.e., the spectral spread of the carrier). A microwave system with a carrier in the 1–10 GHz region might exhibit a linewidth of 1–10 Hz. This gives a carrier frequency to linewidth ratio:

$$\frac{f_c}{\Delta_c} \approx 10^9 \qquad (15.2)$$

For optical systems operating at 1300 nm ($f_c = 231$ THz) typical carrier linewidths and ratios are as follows:

- LED (very low cost) $\Delta_c = 2000$ GHz, $f_c/\Delta_c \approx 10^2$
- FP laser (medium cost) $\Delta_c = 500$ GHz, $f_c/\Delta_c \approx 10^3$
- DFB laser (high cost) $\Delta_c = 20$ MHz, $f_c/\Delta_c \approx 10^7$
- LEC laser (very high cost) $\Delta_c = 100$ kHz, $f_c/\Delta_c \approx 10^9$

So today's leading edge optical systems equal or better their microwave counterparts in these respects. This is a key turning point in optical systems development as it allows channel packing/spectral efficiency comparable with or better than microwave systems, with even longer operating spans through a significant signal-to-noise ratio improvement (Linke, 1989).

From a network point of view, optical systems are following the same broad development and deployment history of microwave radio and copper cables, with an expensive start-up phase of early technology used in long-haul applications (dense traffic) followed by a migration to lower cost systems as the technology matures, then a move towards the short-haul network (thin traffic), and ultimately to customer access (very thin traffic).

15.8 Migration from electronics to optics

In contrast with copper systems, optical fibre has less electronics associated with the transmission path than with the terminal stations. Transmission technology has thus seen a change in the balance of reliability from the basic line plant to the terminal equipments as depicted in Fig. 15.6. Comparing system developments over the last 20 years, it is clear that the terminal electronics is now a dominant feature in the reliability equation (Davidson *et al.*, 1989). Just as the move from FDM to PCM saw improvements in system and network reliability, the move from asynchronous multiplexing AMUX to synchronous multiplexing SMUX (SONET/SDH) offers a similar advance. In the 1990s, optical fibre is able to provide bit rates up to 2.4 Gbit/s over distances greater than 100 km without the need for conventional 3R (reshape, regenerate, and retime) repeaters. As the synchronous digital hierarchy is introduced, there will be a significant reduction in the multiplexing electronics, while providing the option for the switching and add/drop of circuits at $N \times 2$ or $N \times 1.544$ Mbit/s (IEEE, 1990).

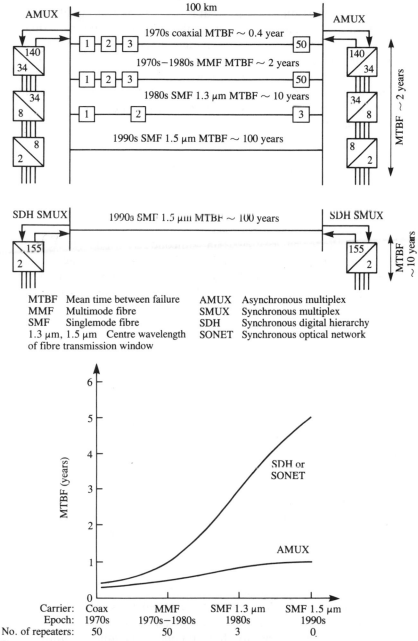

Figure 15.6 Reliability improvements with technology

The hierarchical standards for this scheme have been agreed
internationally; systems are currently being developed and networks
planned (IEEE, 1990).

The obvious question to ask is: what is the next step we could anticipate in this wish list of improvements? It might be supposed that removing the electronic blockage presented by the MUX elements and realizing the full bandwidth potential of the fibre throughout a network would be a reasonable objective. This would be a step towards the transmission engineer's dream—a totally transparent network of infinite bandwidth with minimal hardware and software giving a very high reliability and utility at a very low cost.

15.9 Optical amplifiers

Since 1986, optical amplifiers have emerged as promising network elements (O'Mahony, 1988) for the future, to the point where they are now commercially available for research and trial applications. Both linear and nonlinear devices based on semiconductor and fibre elements have been demonstrated to have stable and controllable characteristics. These devices stand poised to change the face of telecommunications, as they will release the full fibre bandwidth and realize a network that is inherently transparent (Cochrane, 1990).

From an operational standpoint, the optical amplifier offers unparalleled flexibility and utility. It has a demonstrated ability to convey both analogue and digital signals in AM, FM, and OOK, FSK, PSK formats over considerable distances in an FDM/WDM mode. In principle, the mix of signal formats is unimportant—all could coexist in the same fibre and amplifiers at the same time at multi-Gbit/s information rates. Moreover, optical amplifiers can also sustain soliton (perfect wave) propagation (Brown *et al.*, 1991).

An exhaustive comparison of optical amplifiers is not yet possible as many of the fundamental aspects are still being investigated. For example, the present understanding of such topics as the basic noise accumulation mechanisms for cascaded amplifiers, nonlinear distortion, and the ultimate transmission link limitations is still incomplete. However, what is clear from the published results is that linear and nonlinear systems approximately 20 000 km in length are feasible with suitably engineered components. In this context it is possible to draw the following comparisons:

- Travelling wave semiconductor laser amplifiers (TWSLAs) require a similar pump power (electrical) to fibre amplifiers whereas the power required for Raman amplification is substantially higher.
- TWSLAs realize less fibre-to-fibre gain due to the inherent fibre coupling losses, and they are polarization sensitive.
- The very wide bandwidth of TWSLAs and erbium-doped amplifiers promotes a rapid build-up of spontaneous emission noise leading to early gain saturation which may necessitate the use of optical filters on very long distance systems.

- Poor splice and facet reflectivities can drive TWSLAs and erbium amplifiers into a lasing condition and careful control is necessary to maintain system stability.
- Erbium amplifiers so far have the lowest noise and best crosstalk performance of all amplifier types reported. The need for high-power optical pumps may render fibre amplifiers inherently less reliable than TWSLAs, but the use of multilaser pump circuits and/or more efficient dopant/geometry may remove this potential problem.
- The bandwidth of cascaded TWSLAs is dictated by the control of chip characteristics and geometries which define the precise passband and centre wavelength of each amplifier. Temperature, ageing and bias current can also influence these parameters significantly.
- The bandwidth and centre wavelength of fibre amplifiers are mainly defined by the atomic structure and not by mechanical geometry. Variations due to temperature, ageing, and pump conditions are therefore less significant.
- For very long systems (~ 5000 km) there is still some debate about the inherent nonlinearity of fibre and the resulting distortion. Experiments to date have shown pulse compression on systems operating in a WDM mode. In contrast, soliton systems operating in a nonlinear mode have demonstrated an ability to transmit undistorted pulses over distances up to 40 000 km. Soliton systems operating in a non-linear WDM mode are also being investigated, and early results are encouraging.

Photonic amplifiers can thus be expected to be used across a broad front of system applications as indicated in Table 15.3. In broad terms the limiting (signal-to-noise) S/N ratio of an amplified system may be approximated by:

$$S/N = \frac{4P_0}{NG\eta B} \tag{15.3}$$

where P_o = amplifier output power
N = number of amplifiers
G = amplifier gain

Table 15.3 Optical amplifier applications

Application	TWSLA	Erbium (Bidirectional)	Raman	Brillouin (Unidirectional)
Power amplifier	✓	✓	✓	✗
Repeater	✓	✓	✓	✗
Receiver preamp	✓	✓	✗	Narrow band only
Selective filter	✗	✗	✗	✓

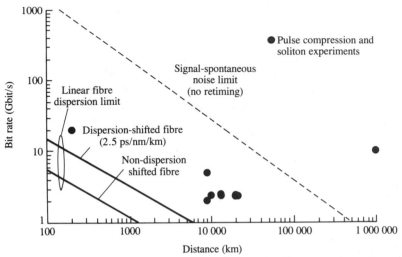

Figure 15.7 Bit rate versus distance limits for linear and nonlinear fibre transmission

B = bit rate

η = spontaneous noise factor

For a BER of 10^{-9}, Eq. (15.3) predicts that $N > 300$ and spans approaching 50 000 km could be realized. However, when fibre dispersion, laser chirp, linewidth, modulation rate, and format are taken into account, this optimistic estimate rapidly becomes more modest as indicated in Fig. 15.7. For intensity-modulated systems, laser chirp and fibre material dispersion broader than the optical pulse can severely limit both the speed and the distance of transmission. Even with coherent optics which remove laser chirp related dispersion, the finite linewidth and spectral spreading due to modulation poses a significant distance limitation (IEEE, 1989; Trischitta and Chen, 1989). Nevertheless, practical demonstrations have now confirmed that transoceanic and transcontinental systems are feasible with standard components. Figs. 15.7 and 15.8 show the reported linear and nonlinear system experiments to date.

15.10 Optical modulation

Optical fibre systems are currently dominated by very simple intensity-modulated schemes with no inherent signal selection at the receiver. Lasers are merely 'turned on and off' by current pulses, and photons are detected directly by APD or PIN diodes. The sudden change in carrier density that results in a laser induces broad dynamic wavelength shifts (referred to as *chirp*) in the emitted light that are both bit pattern

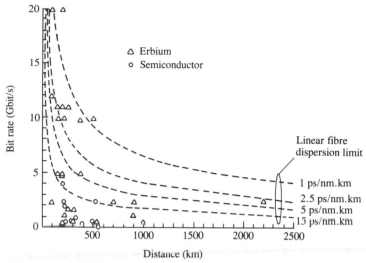

Figure 15.8 Reported multiamplifier experiments and field trials

and rate dependent. This results in a wavelength chirp effect at different rates and points in the bit sequence. For example, a DFB laser with a 50 MHz linewidth at 1535 nm, operating at 2.4 Gbit/s, can typically broaden to 60 GHz as a result of chirping. In turn, this results in pulse dispersion with energy displaced from the zero dispersion wavelength. This is a key limiting mechanism to long-distance transmission, and might typically constrain the span length to 60 km. An obvious solution is the use of constant amplitude FSK format with the laser modulated by a small current deviation to minimize the effects of chirping (Steele, 1989). This then allows the system to operate over 100 km or more. A further option requires an external modulator and optical isolator to allow the laser to operate in an unmodulated carrier wave (CW) mode. This produces a better result, but at the expense of launch power owing to the optical loss of the modulator and, usually unnecessary, isolator. Recent developments in solid state lasers and the requirements of safe operating practice have largely negated this as a problem.

More recently, the need to transport signals at very high bit rates has led to the demonstration of WDM and SCM (sub carrier modulation) schemes, which closely resemble their analogue FDM predecessors (Olshanski, 1988). Assembling such signal stacks is relatively straightforward, but their successful amplitude-modulated transmission relies on a high degree of transmission path linearity. The fact that the majority of the past device development has been concentrated on digital applications does not help as linearity has not been any significant concern. However, the use of analogue FM modulation of the optical source generally overcomes this problem, and signals in this format have been transmitted over 1000 km with 10+ optical amplifiers in cascade at capacities of 16 Gbit/s or more.

15.10.1 The perfect wave

All optical fibres fundamentally exhibit a degree of nonlinearity that, on sub-1000 km systems, generally goes unnoticed as the resulting level of signal distortion is insignificant. However, on very long lengths of fibre this can pose a significant problem. Indeed, calculations show that a 10 000 km route of amplified fibre will see a 50 per cent compression of pulses at 2.4 Gbit/s, although experiments yield somewhat less pessimistic results. Clearly these nonlinear effects cannot be ignored, but they only pose a problem if the system is designed with the assumption that the fibre is a linear transmission medium. If the system design takes into account the intrinsic nonlinearity of the fibre, beneficial use can then be made of this property through the utilization of soliton pulses (Brown *et al.*, 1991)—a form of 'perfect' pulse which allows undistorted transmission over vast distances, well beyond the normal limits governed by linear conditions (Fig. 15.7). Interestingly, recent experimental results suggest that WDM soliton propagation might also be possible. This would perhaps be the ultimate system, with unlimited distance and bandwidth with no signal distortion.

15.11 Future networks

Having removed the need for electronic constrictions in the transmission and multiplex elements through the use of optical transparency and WDM, the question arises as to how we can exploit this new degree of freedom. If we now start to view the bandwidth of fibre in the same way that we consider radio, we rapidly migrate towards an optical ether (Cochrane *et al.*, 1991). Moreover, this is a natural extension of the present PONs being developed for the local loop and LAN applications. The basic configuration of such an ether is shown in Fig. 15.9. To address an endpoint terminal all that is required is the area wavelength (not area code) and the final (fine-grain) wavelength for WDMA, frequency for FDMA, or code in the case of CDMA (this last option is depicted in Fig. 15.9). It has been demonstrated that this approach can be extended to the replacement of the central office for up to 200 000 lines with a 100 per cent nonblocking capability. At today's service levels with a similar level of call/service blocking, then up to 20 000 000 lines could be accommodated. Beyond this level the need for conventional switching of some kind becomes necessary again. Figure 15.10 gives a schematic of a system that has been used to give dimensions to the problem and provide a focus for study. The key assumption is that all the optics and

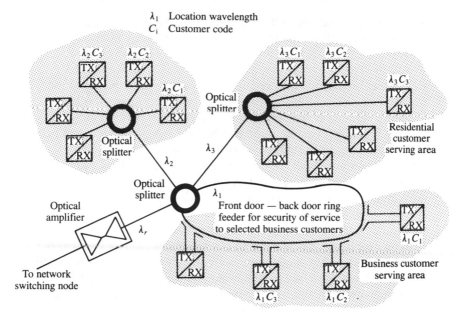

Figure 15.9   The optical ether concept

control for steering and selecting wavelengths resides in the customer
terminal (Cochrane and Brain, 1988).

The ether concept as described has been demonstrated in the
laboratory with a small number of terminals switching 100 Gbit/s
capacities, and in the London fibre network with more modest capacity

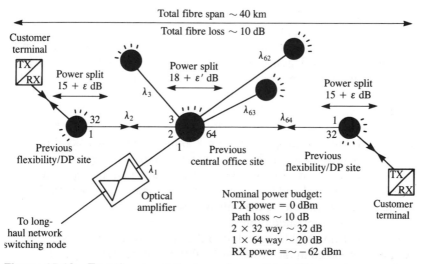

Figure 15.10   Two thousand line passive optical office replacement

levels (Chidgey and Hill, 1989). It works! The key problems and limitations now hinge on the realization of sufficiently miniature and reliable electro-optics, memory capacity, control systems, and wavelength standards. The rate of developmental progress towards all of these requirements currently puts commercial realization in the 2010 time frame. However, on the way, the trajectory of development is likely to see many variants of the same basic concepts. It is perhaps most likely that the first area of exploitation will be for LANs of modest numbers of terminals, where the allocation of an operating wavelength and transparent channel will remove the need for overcomplex and bandwidth-constraining TDMA/ATM schemes. This will become both inevitable and a necessity as computer clock rates exceed 1 GHz and electronic limitations start to impinge above 10 GHz.

Another interesting prospect for transparent networks is that of customer mobility. It is possible to transport raw microwave signals over fibre with only a single stage of frequency translation at each end. A radio cell can therefore be replicated over several remote and diverse locations to give the illusion of a single location. In effect, six company locations start to appear as one to the individual user. The obviousness of this prospect increases with the realization that optical wireless (Cochrane and Todd, 1991) is also being developed for in-building and street use for both telephone and data services.

15.11.1 Equipment practice

The printed circuit board and rack structure have served us well for the past 25 years, but they are now suffering from fundamental limitations that will impact on the density and speed of operation of future systems. High-speed clock and signal distribution is severely limited at rates much above 1 Gbit/s and certainly poses significant problems at 10 Gbit/s and higher. The passive or active optical backplane and card is one solution that is currently under investigation. This uses a glass substrate to act as a transmission path between individual chips and cards. A further alternative is to use D-profile (Fig. 15.11) fibre couplers and amplifying fibre (Cochrane and Todd, 1991; Cassidy *et al.*, 1989) to provide the lossless, infinitely tapped bus, as depicted in Fig. 15.11. The key novelty of this solution is that the bus could start in London, add/drop in New York and terminate in Tokyo! Indeed, the fibre network itself could ultimately be viewed as one vast bus.

So far we have only used two parameters to access data—space and time. There is a third and highly valuable parameter—wavelength. Using this additional parameter is the only way we can increase the bits per unit-volume of accommodation without exceeding the energy per unit-volume of existing building stock into the new millennium. This is elegantly embodied in the powerful MONET concept (Healey *et al.*, 1990) a schematic of which is shown in Fig. 15.12.

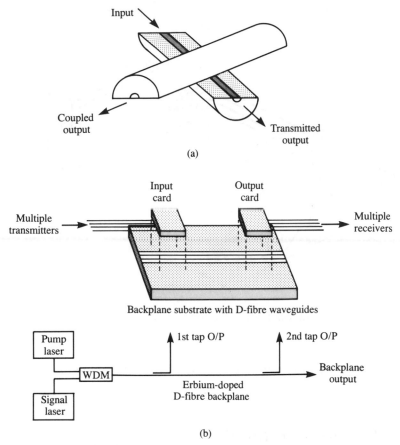

Figure 15.11   (a) Principle of D-fibre coupler; (b) amplified optical backplane/bus using erbium-doped D-fibres

15.11.2 New services

The storage and processing power of computers continues to grow exponentially, as does telecommunication capacity through the demand for mobility and data services of all kinds (Palfreman and Swade, 1991; Engleberger, 1989). As we reach the year 2000, the cost and limits on travel, robotics, artificial intelligence, telepresence, telegames, telemedicine, independent machine–machine communication, and many other new services will maintain the demand for bandwidth and time (ITU, 1991). Distance and bandwidth will have become irrelevant in the new millennium, and service plus time will have become the dominant parameters for charging. We can also anticipate the migration of the local call from 'just London' to the whole of the UK, then all of Europe, and ultimately the world! This can only be so if optical

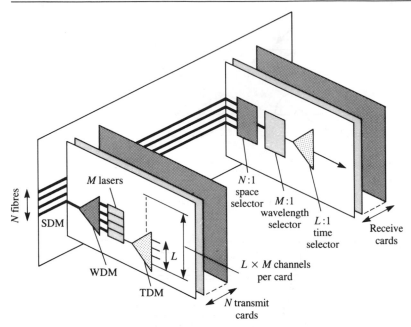

N fibres

SDM

M lasers

WDM

TDM

L

N transmit cards

N :1 space selector

M :1 wavelength selector

L :1 time selector

L × *M* channels per card

Receive cards

Figure 15.12   The MONET principle for *L* × *M* × *N* channels

amplifiers and network transparency have become ubiquitous and customers can gain direct access to networks.

15.12 Final comments

Much of today's development effort is directed towards a single unified network based upon a single open architecture to meet all communication needs. With the present strain of technology this is likely to remain prohibitively complex and expensive, even to the point of impossibility. Not only is the electronics required excessive but the associated software will introduce orders of magnitude of additional risk and inherent unreliability. The arrival of the optical amplifier and transparent optical network (TON) is therefore timely and, indeed, essential if we are to maintain the present rate of progress and continue to satisfy expanding customer service requirements. To do this we must also break away from the established 'mind set' put in place by the earlier copper technology. TONs will open up a series of new avenues for development that will drastically reduce the network management burden (IEE, 1988) and change the balance of switching, transmission, signalling, control, customer equipment complexity, regulation, and competition. In short—a complete paradigmatic change!

References

Barty-King, H. (1979) *Girdle Round The Earth*, Heinemann, Stoneham, Maryland.

British Telecom (1989) 'The Local Network', *British Telecom Technology Journal*, special issue, vol. 17, no. 2.

Bronowski, J. (1973) *The Ascent of Man*, BBC Books, London.

Brooks, R. M. (1980) 'Line coding for optical fibre submarine systems', M.Sc. Dissertation, University of Essex.

Brown, G. N. *et al.* (1991) 'Computer simulation of ultra-high capacity non-linear lightwave transmission systems', *3rd IEE Conference on Telecommunications*, Edinburgh, pp. 319–23.

Bylanski, P. and Ingram, D. G. W. (1980) *Digital Transmission Systems*, Peter Peregrinus, London.

Campbell, G. A. (1937) Collected Papers, AT&T, New York.

Cassidy, S. A. *et al.* (1989) 'Extendible optical interconnection bus fabricated using D-fibres', *IOOC '89*, Kobe, Japan, paper 21D2-1, pp. 88–9.

Cattermole, K. W. (1969) *Principles of Pulse Code Modulaton*, Iliffe, London.

Chidgey, P. J. and Hill, G. R. (1989a) *IEE Electronics Letters*, vol. 25, no. 21, 1451–2.

Chidgey, P. J. and Hill, G. R. (1989b) 'Wavelength routing for long haul networks', *ICC-89*, Boston, p. 23.3.

Cleick, J. (1987) *Chaos*, Cardinal.

Cochrane, P. (1984) 'Future trends in telecommunications transmission—a personal view', *Proc. IEE*, F 131/7, 669–83.

Cochrane, P. (1990) 'Future directions in long haul fibre optic systems', *British Telecom Technology Journal*, 8/2, 5–17.

Cochrane, P. and Brain, M. C. (1988) 'Future optical fibre technology and networks', *IEEE COMSOC Mag*, 26/11, 45–60.

Cochrane, P., Heatley, D. J. T. and Todd, C. J. (1991) 'Towards the transparent optical network', *ITU Word Telecommunications Conference*, Geneva, pp. 105–9.

Cochrane, P. and Todd, C. J. (1991) 'Towards network transparency with photonic amplifiers', *OSA Conference on Optical Amplifiers and their Applications*, Snowmass, CO, *OSA Digest*, vol. 13, WA1/1–5.

Cox, S. A. and Adams, P. F. (1985) 'An analysis of digital transmission techniques for the local network', *British Telecom Technology Journal*, 3/3, 73–85.

Crank, G. J. and Cochrane, P. (1980) 'High speed digital transmission in the British Post Office coaxial network', *POEEJ*, 73/3, 145–52.

Davidson, J., Hawker, I. and Cochrane, P. (1989) 'The evolution of service protection in the BT network', *IEEE Globecom 89*, Dallas, Nov. 1989, p. 23.6.1.

Engleberger, J. (1989) *Robots in Service*, Kogan Page, New York.

Flood, J. E. and Cochrane, P. (1991) *Transmissions Systems*, Peter Peregrinus, London.

Griffiths, J. M. (1983) *Local Telecommunications*, Peter Peregrinus, London.

Griffiths, J. M. (1990) *ISDN Explained*, John Wiley, London.

Healey, P., Cassidy, S. and Smith, D. W. (1990) 'Multidimensional optical interconnection networks', *Proc. SPIE* (Digital Optical Computing II), Int. Soc. Opt. Eng. (USA), vol. 1215, 191–7.

Hughes, C. J. (1986) 'Switching—state of the art', *British Telecom Technology Journal*, vol. 4, nos. 1 and 2.

IEE (1984) IEE Colloquium on Interference and Crosstalk on Cable Systems, London.

IEEE (1988) 'Telecom network operations and management', *Journal on Selected Areas in Communications*, vol. 6, no. 4.

IEE (1989) IEE colloquium on optical amplifiers for communications, London.

IEEE (1986) 'Engineering and field experience with fibre optic systems', *J. SAC*, special issue, 4/5.

IEEE (1989) 'Telecommunication deregulation', *IEEE Communications Magazine*, special issue, 27/1.

IEEE (1990) 'Global deployment of SDH compliant networks', *IEEE Communications Magazine*, special issue, vol. 28, no. 8.

IEEE (1991) 'The 21st century subscriber loop', *IEEE Communications Magazine*, special issue, 29/3.

ITU (1991) *World Communications—Going Global With a Networked Society*.

Linke, R. A. (1989) 'Optical heterodyne communication systems', *IEEE Communications Magazine*, vol. 27, no. 10, 36–41.

Nahin, P. J. (1987) *Oliver Heaviside—Sage in Solitude*, IEEE Press, Piscataway, New Jersey.

Olshanski, R. (1988) 'Sixty channel FM video SCM optical communication system', *IEEE Optical Fibre Conference (OFC)*, p. 192.

O'Mahony, M. J. (1988) 'Semiconductor laser amplifiers for future fibre systems', *IEEE Journal of Lightwave Technology*, vol. 6, no. 4, 531–44.

Palfreman, J. and Swade, D. (1991) *The Dream Machine*, BBC Books, London.

Reeves, H. S. V. (1978) '140 Mbit/s digital line system for coaxial cable', *Electrical Communication*, vol. 53/2, 173.

Steele, R. C. (1989) 'Recent progress in coherent systems research at BT', *Proc. SPIE Opto Electronics and Fibres*, Boston, p. 1175.

POEEJ, Special Issue on 60 MHz Transmission System, 66/3, 1973.

Trischitta, R. and Chen, D. T. S. (1989) 'Repeaterless undersea lightwave systems', *IEEE Communications Magazine*, vol. 27, no. 3, 16–21.

Main abbreviations

2B1Q	two binary to one quaternary
AT&T	American Telephone and Telegraph Co.
ADM	add-drop multiplexer
ADPCM	adaptive differential pulse code modulation
AMI	alternate mark inversion
ANSI	American National Standards Institute
APD	avalanche photodiode
ARQ	automatic request for retransmission
ASK	amplitude shift keying
ATM	asynchronous transfer mode
B-ISDN	broadband integrated services digital network
BCH	Bose-Chaudhuri-Hochquenghem
BER	bit-error ratio
BR	basic rate
BWTD	bandwidth times distance
CBR	constant bit rate
CCIR	International Radio Consultative Committee
CCITT	International Consultative Committee on Telephony and Telegraphy
CIF	common intermediate format
CLP	cell loss priority
CMI	code mark inversion
CRC	cyclic redundancy check
DFB	distributed feedback
DLE	digital local exchange
DMSU	digital main switching unit
DP	distribution point
DSF	dispersion-shifted fibre
DSI	digital speech interpolation
DSP	digital signal processing
EFS	error free seconds
EHF	extra high frequency
ES	errored seconds
FAW	frame alignment word
FDM	frequency division multiplex
FEC	forward error correction
FP	Fabry–Perot
FSK	frequency shift keying
GSM	Groupe Speciale Mobile

HDTV	high-definition television
i.s.i.	intersymbol interference
IDN	integrated digital network
IF	intermediate frequency
ISDN	integrated services digital network
ITU	International Telecommunication Union
LAN	local area network
LTE	local telephone exchange
MAN	metropolitan area network
MAP	maximum aposteriori probability
ML	maximum likelihood
MMF	multimode fibre
MTBF	mean time between failures
MUX	multiplexer
NA	numerical aperture
NTE	network terminating equipment
NTT	Nippon Telephone and Telegraph
OAM	operations, administration and maintenance
OFCS	optical fibre communication system
OOK	on–off keying
PABX	private automatic branch exchange
PAM	pulse amplitude modulation
PCM	pulse code modulation
PDH	plesiochronous digital hierarchy
PON	passive optical network
PR	primary rate
PSK	phase shift keying
PSTN	public switched telephone network
Q	quality factor
QAM	quadrature amplitude modulation
QPSK	quadrature phase shift keying
RCU	remote concentrator unit
RF	radio frequency
S/N	signal-to-noise
SDH	synchronous digital hierarchy
SES	severely errored seconds
SHDTV	super high-definition television
SMF	singlemode fibre
SNR	signal-to-noise ratio
SONET	synchronous optical network
STM	synchronous transport module
TDM	time division multiplexing
TPON	telephony passive optical network
UHF	ultra high frequency
UNI	user–network interface

VBR	variable bit rate
VHF	very high frequency
VLSI	very large scale integration
WDM	wavelength division multiplexing
VBR	variable bit rate

Index